D0580018

From Edison to iPod

PROTECT YOUR IDEAS AND MAKE MONEY

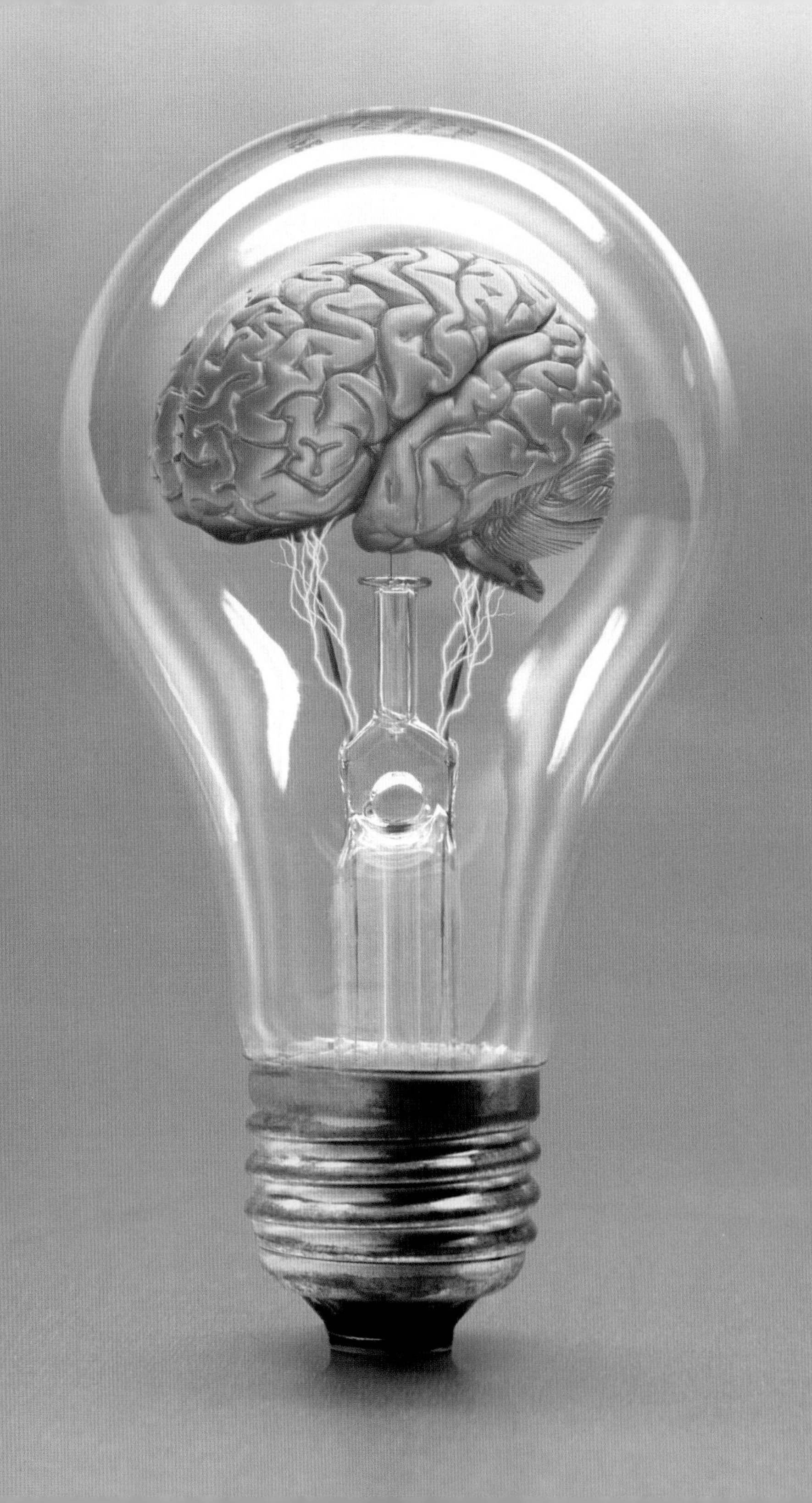

FREDERICK W. MOSTERT
AND LAWRENCE E. APOLZON

From Edison to iPod

PROTECT YOUR IDEAS AND MAKE MONEY

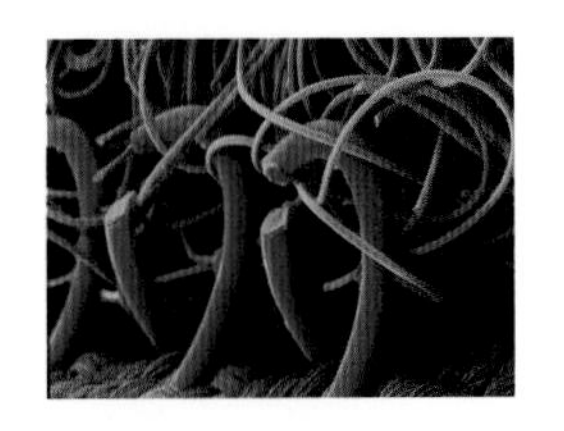

LONDON, NEW YORK, MUNICH, MELBOURNE, DELHI

IN MEMORY OF THAT REMARKABLE JUDGE, ANTON W. MOSTERT,
AND LARRY'S LOVING PARENTS

Editor Elizabeth Watson
Senior Art Editor Helen Spencer
Executive Managing Editor Adèle Hayward
Managing Art Editor Nick Harris
DTP Designer Traci Salter
Production Controller Luca Frassinetti
Art Director Peter Luff
Publishing Director Corinne Roberts
Designed for DK by Bill Mason

First published in the United States in 2007 by
DK Publishing, Inc.
375 Hudson Street
New York, NY 10014

A Penguin Company

07 08 09 10 11 10 9 8 7 6 5 4 3 2 1

A Cataloging-in-Publication record for this book is available from the Library of Congress.

ISBN: 978-0-7566-2602-0

Reproduced by MDP in the UK
Printed and bound in China by Leo Paper Products Ltd
Discover more at www.dk.com

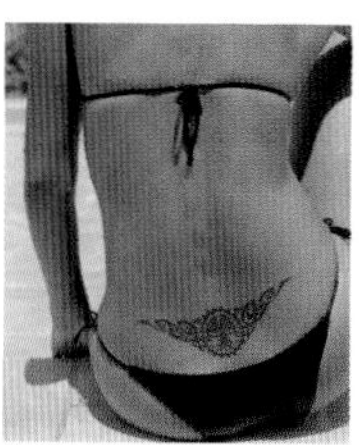

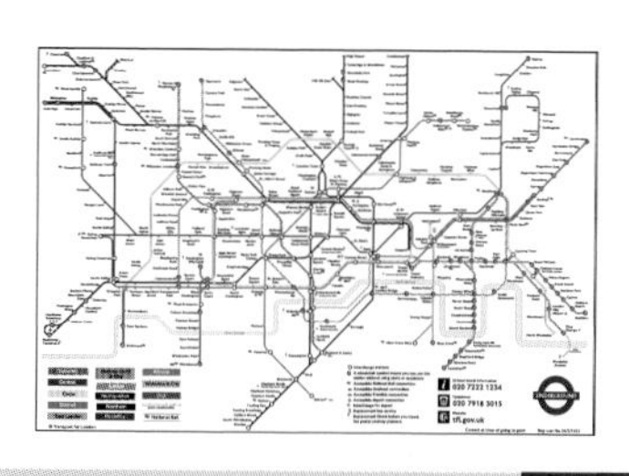

DISCLAIMER

All information in this book is provided to the public as a source of general information on intellectual property issues. In legal matters, no publication — whether in printed or electronic form—can take the place of professional advice given with full knowledge of the specific facts and laws that apply. While reasonable effort has been made to ensure the accuracy of the information in this book, it should be treated as the starting point for understanding how intellectual property can be used to help you and your business and how it fits into the creative process. It should be emphasized that intellectual property laws and the way they are interpreted vary from country to country, and even between the states in the US and the various courts. The information included in this book will not be relevant or accurate in other countries. The examples and references to intellectual property in this book do not constitute any judgement or view by the publisher or authors as to the validity or enforceability, or lack thereof, and are intended for educational purposes only.

KEY TO SYMBOLS

! Note of caution: something to be aware of

Definition: a jargon-busting explanation of legal terms

Debunking the myth: sets straight common misbeliefs

Legal test: factors the law takes into account when addressing an issue

Good news!

While you should always work with your lawyer, this is a special reminder

Useful resources

? A whole-page feature addressing a particular concern

Indicates a list of examples

contents

You can also find more information at www.fromedisontoipod.com

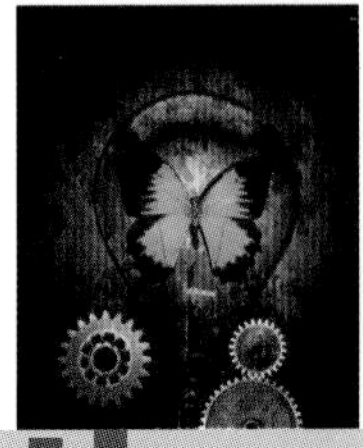

introduction

Intellectual property has become a household issue. From the boardroom to Internet chat rooms, and even in gossip columns, intellectual property is a hot topic. But what exactly is it? How does it affect you? If you make it big, isn't this something your lawyers can just take care of? Or you might modestly conclude that you are not an intellectual, you do not have any property, and therefore there is no way that understanding intellectual property would matter to you.

It matters! Understanding intellectual property can help you protect an incredible creation and parlay it into a valuable asset that you can exploit, license, or sell. Without a grasp of intellectual property, you may unwittingly give your creation away. Adopting a few safeguards can make the difference between building a business and regretting missed opportunities. And when intellectual property is approached as a whole, it becomes a reliable tool available to you and your business: a resource you can put to work.

If you are a creative person, like a designer, an artist or a musician, an entrepreneur with unique concepts, or a marketer, or if you are an investor who is frequently approached with new ideas, this is a good guidebook.

The Internet and other technological advances have given our generation immediate access to endless information. Corporate downsizing has introduced more individual entrepreneurs and consultants to the marketplace and the global economy makes it essential to have knowledge and tools available to remain competitive. Cultural and business climates have evolved to a point where many people are sophisticated enough to know their creation is entitled to valuable intellectual property rights. But you may be unsure as to how to protect these rights. What's the difference between a utility patent and a design patent? A trademark and copyright? What's a trade secret, and how does it differ from a trade name? What about domain names? These categories might sound like far-off places and all you have is a blurry out-of-date road map to find your way. But when all these areas of intellectual property are understood and approached holistically, you will have the power to make good business decisions and get the most out of your creations.

Since the 1980s, we have watched an interest in and awareness of intellectual property skyrocket. At social and family gatherings, we have gone from watching eyes glaze over as we explain what we do to being asked very specific questions about a particular issue someone is facing. These days we attend social gatherings armed with pens and notepads because we are often cornered by someone looking for some "quick" advice on how to protect a new creation. For example, how to protect a recipe for a new pastry creation, a new website design, an idea for an innovative bracelet, or a new restaurant name. We have advised a wide variety of individuals, including watch and jewelry designers, chefs, comedians, physical fitness trainers, opera singers, restaurateurs, computer scientists, actors, fashion designers, record labels, doctors, fragrance houses, movie producers, architects, corporate finance specialists, pop artists, symphony orchestras, bankers, and numerous other entrepreneurs. Our work has included looking after the interests of not-for-profit organizations and charities as well as working with and advising celebrities and public figures, including Nelson Mandela, Michael Douglas, Catherine Zeta-Jones, Jackie Chan, Sylvester Stallone, Ernie Els, Boris Becker, Stella McCartney, and the Shaolin Monks. However divergent their interests, all these people share one common characteristic—a creative spirit. Not only have they made use of their imaginations, they also possess enthusiasm, optimism, and pride in their creations. They come to us to help them get the

right kind of protection. For us, the joy of intellectual property is the fact that we deal with the creative initiatives of people who come up with excellent ideas. Indeed, we are exposed to the cutting edge of intellectual creation, which can often be converted to useful commercial gain.

Given these needs, we have written this practical guide to demystify intellectual property. Our goal is to reveal the secrets and explain the complexities and formalities of the law, steering clear of legal jargon. We have mapped out the array of intellectual property options available to you and explained how you can develop strategies and practices to protect your creations. We have also included hands-on tips and deeper insights based on many years of advising clients in this field. We like to see ourselves as tour guides taking you on a trip through the world of intellectual property. While the different categories—trademarks, copyrights, trade secrets, etc.—are destinations in themselves, as any good traveler knows, you need to take in the local color to really understand a place. It would be a shame to walk away from this book thinking that intellectual property is a bunch of disjointed, unrelated fields. Don't treat it like, "If it's Tuesday, I must be in utility patents." And, like any good trip, a second visit will bring better understanding. First, read this book in its entirety to see how you can start to apply it in your day-to-day business. Then re-read the sections that are most relevant right now. Finally, re-read the other sections to see whether they may apply later on. But remember, each case depends on its individual facts. Intellectual property laws are filled with nuances and exceptions to the rule. While you may want to try some of this on your own, we can't stress enough how important it is to have a trusted intellectual property lawyer who can advise and guide you through the varying terrain.

When know-how and creativity come together and you manage to create something of intellectual value, it is a great achievement. Well done. Now let's get down to the business of making sure the rights to your brainchild are protected.

the basics

THE GOLDEN RULES

There are two golden rules that will assist you greatly in understanding the fundamentals of intellectual property rights.

Rule number one: first in time, first in right

The one who is first past the post—whether it is tortoise or rabbit—wins. This holds true in fairy tales and in real life. In the past, for example, water laws operated on the same basis: the individual who first actively diverted water from a stream for beneficial use had the first right to use the water.

If you are the first person to file a patent application on a new invention, such as a uniquely engineered safe-seal bottle top or a design patent application on a uniquely designed lamp, or if you are first to use a trademark such as a cool name or hip logo for your product or service, or first to write down words for a new song that have been going through your head, or you have designed a new website in source code, you are in a far better position to protect the intellectual property in your creation and to win any legal battles along the way.

In this guide we will show you how best to employ practical tips to your advantage. They should enable you to demonstrate convincingly that you were the first to establish a particular intellectual property right.

Rule number two: ideas alone are not protectable

We all have ideas and dreams. But that doesn't put money in the bank. When you come up with a specific idea, lay it out and do something with it. The result might be a painting, a play, an improved way to use debit cards on the Internet, a new name for your jewelry line, or a great design for a cell phone. All of these are "creations," which can be protected by different and overlapping categories of intellectual property. But only if you get your idea out of your head and into action.

WHY BOTHER PROTECTING INTELLECTUAL PROPERTY?

This book is not intended to be a history lesson to explain how and why intellectual property came about or how it is evolving. These are very interesting topics and, in fact, at some point you may want to read more about them. But our purpose here is to get you to focus on how you can protect your creations. To do that, we will spend a moment to explain that, in the US and many other countries, the driving force behind intellectual property is that society wants to reward and provide incentives to creative individuals who come up with new creations. This isn't because our society

is a great humanitarian culture, although it may be. The reason here is economic. Using candles and kerosene at nighttime might sound quaint, but it is a big hassle. It starts fires. And it is not very effective, unless your goal is romance. So people like Thomas Edison came along, and we know how that story goes.

Society wants to encourage more than patentable inventions. It provides copyright protection to the authors and visual artists of our time who can collect royalties for their works. The same is true in the commercial world with product and package designs. Design patents are available to protect the fruits of market studies and product research. They enable owners to carve out special niches so that designs can be exploited exclusively. Trademarks are similarly protected on the basis of perceived societal benefit. Consumers need to be safeguarded against deception and confusion by counterfeiters and those who falsely trade on the name and reputation of the original owners. It is in this tradition of providing incentives to create that most of the intellectual property laws have been written. You can be assured that both government and society encourage you to be innovative and reward those who are first to create something of value. Most intellectual property rights are protected on the premise that society as a whole benefits from new scientific inventions and artistic works of creative individuals. Creative spirits, like you, should be incentivized to continue to produce works for the betterment of the community. The best way to accomplish this goal is to grant creative individuals, such as you, exclusive rights to their creations for limited periods of time. Your creations, whatever they are, need and deserve protection. Although free competition in the marketplace is encouraged, it should be fair. Consequently, the law is designed to protect you from unconscionable competitors who steal your ideas. For example, the brand name of your product can be protected by using it or by filing a trademark application, your new invention of a computer-coded wizard to authenticate online sales can be safeguarded by a patent, your pastry recipe can be shielded by trade secret law, your distinctive perfume bottle can be covered by a design patent, and your original design for a unique website or sculpture can be protected by copyright.

The other side of the coin

The idea that you are entitled to benefit from your creative endeavors and that such rights should be yours in perpetuity seems only fair and stems from a basic sense of natural justice. But little in life is black and white. And the world of intellectual property is no exception. Just as society rewards you for your creations, it also needs to draw lines and limit the boundaries of your rights. Free speech, including the right to know, needs to be considered when the media broadcast news items. And handing over inventions to the public for their own use and development comes at the end of a patent.

Trademarks can't be like unused toys in the playground. Either you use them or they are there for someone else to use.

In a free and idea-rich society such as ours, creative spirits need to be nurtured and encouraged so that the creative process can prosper. No man is an island, and we rely on each other for insight and inspiration. Creative spirits must be able to draw on prior inventions and artistic works in order to reach even greater heights. Even such a genius as Sir Isaac Newton admitted with humility, "If I have seen further, it is by standing on the shoulders of giants."

A creative spirit thrives on past experiences, prior inventions, and previously experienced artistic works. This is the way our civilization has advanced for years, and nothing we have seen in over two decades of experience has caused us to believe that it is changing. In fact, the more new ideas emerge into creations, the more exciting and endless it seems. We have watched the Internet unfold and develop from a simple communications jalopy to an entire platform and medium for living and doing business. What will be next? Perhaps one of you reading this book knows. In fact, we bet that a number of you might have the next big idea, and we want to help you navigate your way through the world of intellectual property so that, as your ideas evolve into creations, you will know which types of intellectual property rights you have and how they can be protected.

trademarks

DEFINITION

Trademarks are words, designs, and other markings that identify and distinguish products or services from one another. They are the link between the maker and the customer and they help consumers to distinguish between competing brands—for example, think **COKE** versus **PEPSI**. Trademarks can suggest a certain quality or status (think **ROLLS-ROYCE** versus **MINI**) and they serve the public interest by protecting consumers from being deceived or misled as to a product's real origin.

DURATION

Indefinite, as long as the trademark is used. Since protection flows from use, you have to use them or you lose them. Trademarks often become your most valuable intellectual property after other assets have expired.

◀ **Logo marks**

The iconic Apple logo symbolizes the harmony of excellence in design and computer science. This Apple logo must be one of the most recognized logos throughout the world. The clean design of the logo itself underscores the virtue of simplicity in design.

Trademarks are words, designs, and other markings that identify and distinguish products or services from one another.

Kodak

◄ Word marks
The word Kodak was registered as a trademark in 1888 and was "invented" by George Eastman, "I devised the name myself. The letter 'K' had been a favorite with me—it seems a strong, incisive sort of letter. It became a question of trying out a great number of combinations of letters that made words starting and ending with 'K'. The word 'Kodak' is the result."

► Logo marks
This logo was given to the monks of the Shaolin Temple by a prince of the Ming Dynasty around 1500 AD. It symbolizes the Temple's harmonic fusion of Taoism, Confucianism, and Buddhism. The Shaolin Temple has since stylized this ancient symbol to reflect the Temple's heritage and history. The first character, looking directly at you, represents Buddhism, while the two characters who look at each other, on the left and right of the circle, represent Confucianism and Taoism. This logo can be used only by those who have a direct connection to the Temple—this is an interesting case of trademark law protecting cultural values and heritage.

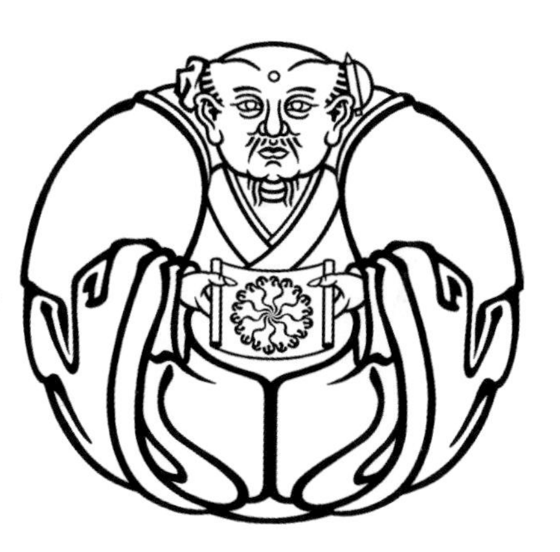

◄ Logo marks
The subject of a trademark may bear no logical relation to the particular product on which it is used, but can become closely associated with that specific product over time. This Penguin logo has become so internationally synonymous with books that, while in solitary confinement in Beirut, hostage Terry Waite drew a picture of a penguin to communicate to his guard his request for some good reading material. And he was understood!

Because protection for trademarks flows from use, you have to use them or you lose them.

HOW TO PROTECT IT

- Through use on your product or service.
- Seriously consider registration with the US Patent and Trademark Office.

TIPS

- Select a strong mark.
- Keep records of the use of the mark, especially the date of first use.
- Obtain a trademark search before use or filing to avoid disappointment and to evaluate risk.
- Consider adopting the same domain name or trade name as your trademark.
- Police and enforce your trademark.
- Use ® for registered and ™ for unregistered trademarks as a warning to others.

▲ **Logo marks**
"Shell" was first used as a trademark by Marcus Samuel and Company in 1891 for kerosene shipped from London to East Asia. This small business originally sold antiques, curios, and, most notably, oriental seashells. The first logo, a mussel shell, was introduced in 1901 and morphed into a scallop shell or "pecten" logo in 1904 and has remained the symbol of the Royal Dutch Shell Group ever since.

EXAMPLES

HEINZ, **IPOD**, the **NBC** peacock logo, the **COCA-COLA** bottle, the shape of the **ROLLS-ROYCE** hood, **DHL**, the name **TINA TURNER**, the color of **VEUVE CLICQUOT** labels, the three rings at the center of a **MONTBLANC** pen.

◀ Combination marks

Heinz is a classic trademark, consisting of an internationally known name, "Heinz," and a logo. Heinz is famous for its "57 Varieties" slogan in the US. It was founded in 1869 in Sharpsburg, Pennsylvania, by Henry John Heinz and its first products were processed condiments delivered to local greengrocers by horse-drawn carriages. In 1876, the company launched its most famous product: tomato ketchup. And the rest is history.

▶ The Michelin Man

Bibendum, or the Michelin Man, has been the key symbol of the French tire company since 1898. It is said that Mr O'Galop, the creator of this logo, derived his inspiration from a stack of rubber tires. A highly active *bon vivant*, Bibendum's feel-good energy is used as a marketing tool. The famous art nouveau Michelin building in Chelsea, London, houses a well-known restaurant under the name of Bibendum, which pays homage to the character Bibendum to this day.

copyright

DEFINITION

Copyright covers creative expression in the broadest sense. The moment you use your own ingenuity to create an original story, poem, article, or even software, you automatically have copyright in your creation. However, in order to qualify, a work must have been rendered in a tangible medium: only the physical expression of an idea is protected, not the idea itself. As well as an economic right, some copyright owners also have moral rights. These protect the way in which your work is used and attributed to you.

DURATION

For works created now, for the life of the individual creator plus 70 years or, for works created by employees for their employer or works for hire, plus 95 years from publication.

Registered User No. 06/E/1655

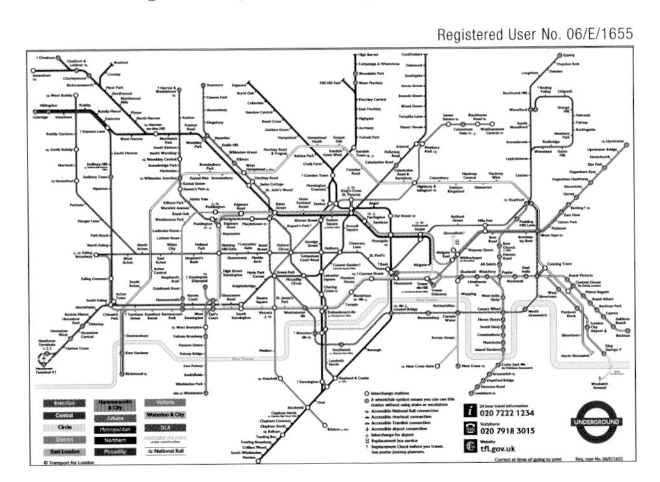

◀ **Maps**
In 1933, Harry Beck, an electrical draftsman, persuaded London Transport to abandon its traditional, free-flowing map and use a diagrammatic map in its place. Beck's training as an electrical engineer led to a vertical and horizontal map with 45-degree angles that is supremely user-friendly and has become a template for transit maps throughout the world.

The moment you use your own ingenuity to create an original story, poem, article, or even software, you have copyright in your creation.

▲ **Sculptures**
The Kiss by Auguste Rodin (1901–1904) has become ubiquitous worldwide as a symbol of love. This masterpiece depicts the lovers Paulo Malatesta and Francesca da Rimini, who were slain by Francesca's incensed husband and immortalized in Dante's *Inferno*.

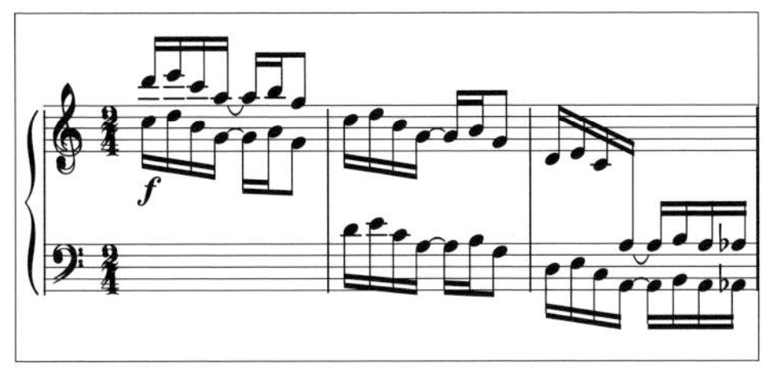

▲ **Musical scores**
The Entertainer, written in 1902 by ragtime musician Scott Joplin, conjures up images of another era, but has found new life as a novelty ringtone. It was also used as the theme music for the 1973 motion picture, *The Sting*.

► **Paintings**
The *Mona Lisa* (1503–1506), with her enigmatic smile, is often named as the most instantly recognizable piece of art in the world. Leonardo da Vinci's famous lady has her home in the Louvre in Paris. Leonardo was so fond of this painting that he is said to have carried it with him wherever he went.

HOW TO PROTECT IT

- It is automatic on the date your creation goes into a tangible medium.
- Consider obtaining a registration through the US Copyright Office to assist proof of ownership and make automatic damages and attorneys' fees available if copied.

TIPS

- Date and sign or stamp to show proof of ownership.
- Consider obtaining notary or third-party confirmation of date of creation or an affidavit.
- Use a copyright legend as a warning: © 2007 John Smith. All rights reserved.

▲ **Tattoos**
Even tattoos can attract copyright if the design is original and unique. However, the person who has the design tattooed on his or her body may not be the owner of the copyright. That right belongs to the original tattoo artist.

EXAMPLES

Literature, writings, paintings, artwork, musical compositions, computer software and source code, visual designs, logos, photographs, jewelry designs, architectural drawings, the design of a web page, radio and television broadcasts, motion pictures (including DVDs), sound recordings of performances (including downloads), fabric patterns, play scripts, sculptures, articles written for magazines, journals, etc., engineering drawings.

As a copyright owner, you also have moral rights. These protect the way in which your work is used and attributed to you.

► Intricate jewelry
The jewelry piece shown here depicts an elegant snake and is an example of the type of unique and elaborate jewelry designs that can be protected by copyright. Obtaining copyright protection is an effective way of building up your arsenal against cheap knock-offs.

RIGHTS OF PUBLICITY

In the United States, another right that can be protected is known as the "right of publicity." This protects the name, likeness, voice, and other characteristics of celebrities and people who have "selling power." This right can last for as long as 100 years after death. Even though these rights exist and can be very useful in the United States, if you plan to use or license your name as a logo with your image or some other trait you have, for specific products and services, it is also a good idea to file a trademark application and get the necessary domain name registrations before you become too popular. Outside the United States, where this "right of publicity" is not recognized, obtaining trademark registrations and domain name registrations is your best option.

design patents

DEFINITION

Design patents protect the two-dimensional and three-dimensional appearance or ornamental shape of a product or its packaging. To qualify, an object must have a specific appearance that can be visually recognized; a design patent has nothing to do with the task an item performs. If the item functions in a particularly unique way or performs a unique task, you should also consider protecting it as a utility patent.

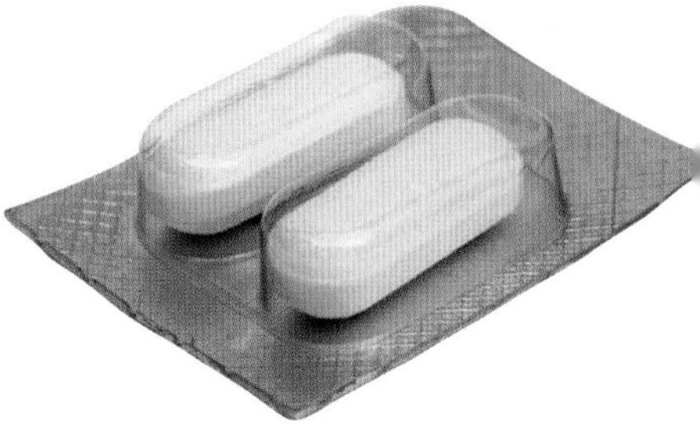

▲ **Package design**
Like any type of packaging, blister packs for pills or other items can be worthy of design patent protection. With so much potential variation, it is no surprise that dozens of designs have been patented for packaging worldwide.

DURATION

Fourteen years after the design patent issues.

Design patents protect the two- or three-dimensional appearance or ornamental shape of a product or its packaging.

► **Designer handbags**
One of the key distinguishing features of the iconic Chloe bag is the embroidery inspired by a vintage bag bought in a flea market. It provides a feminine touch and gives the bag its "Chloe Spirit." The design process for this bag was unusual in that it took only one month from first drawing through to final prototype.

▼ **Automobile designs**
The E-type Jaguar (1961) is considered one of the sexiest sports cars ever launched and the crowning glory of Sir William Lyons, founder of the Jaguar motor company. A take on its racing sibling, the D-type Jaguar, the E-type won the Le Mans 24-hour endurance race five times. The shape was the brainchild of ex-aircraft-industry specialist Malcolm Sayer, who combined artistic flair with wind-flow dynamics when designing the car's contours. The E-type Jaguar is often the only automobile on display at the Museum of Modern Art in New York.

Consider a patent search before filing to avoid disappointment or huge expenses.

HOW TO PROTECT IT

- File a design patent application with the US Patent and Trademark Office.
- This must be done within one year of first offer for sale and public activities and six months after any foreign filings if they issue during that period.

TIPS

- Run to the US Patent and Trademark Office to beat the competition.
- Before that, keep good records, and date them.
- Consider a patent search before filing to avoid disappointment or huge expenses.
- Don't disclose your invention before filing!
- Have business partners, subcontractors, and employees sign confidentiality letters or nondisclosure agreements until you are on file.

EXAMPLES

Perfume bottles, jewelry, watches, car bodies, computer icons, and product packaging.

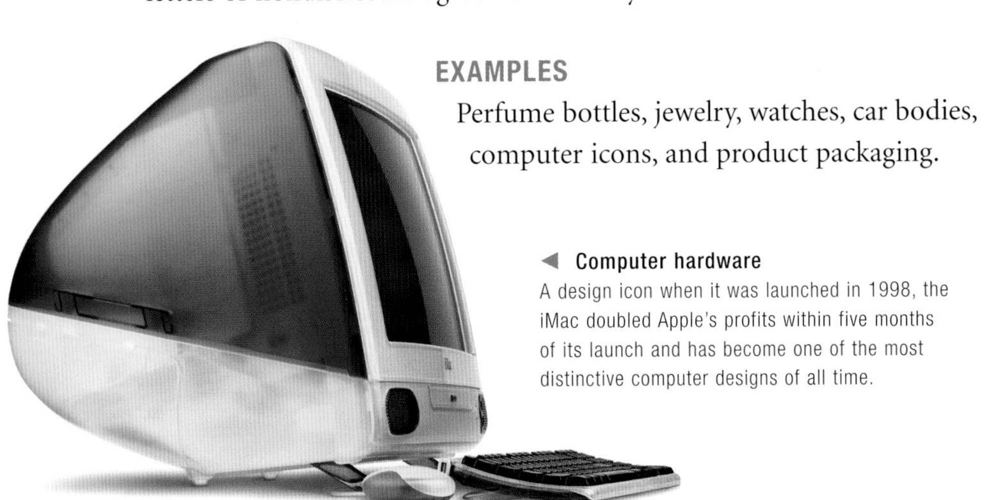

◄ **Computer hardware**
A design icon when it was launched in 1998, the iMac doubled Apple's profits within five months of its launch and has become one of the most distinctive computer designs of all time.

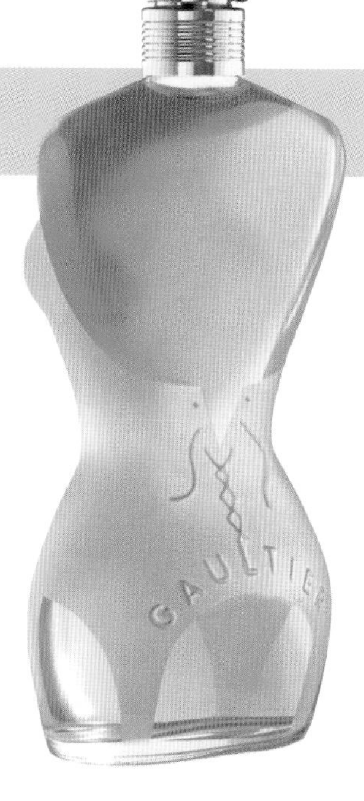

▲ Perfume bottles
This unique, distinctive perfume bottle design by Jean-Paul Gaultier has become a classic example of design in the perfume world. In designing this icon, Jean-Paul Gaultier was inspired by Madonna wearing her trademark bustier. The design is completed by a powdered glass dusting.

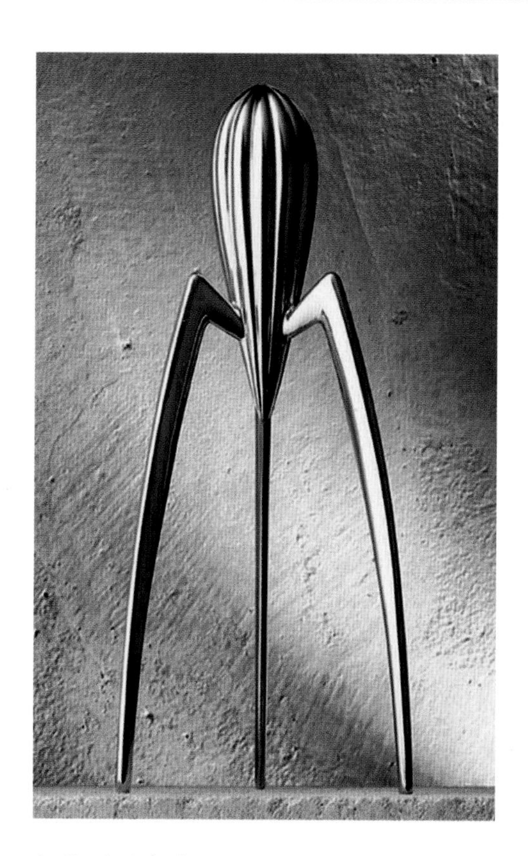

▲ Product design
This lemon squeezer was designed by Philippe Starck in 1990. Bizarrely, the product is now more famous as a design icon than as a useful kitchen implement. This product proves the point that everyday objects can become true design icons.

trade secrets

DEFINITION

Trade secrets refer to confidential business information, such as formulas, technical know-how, computer code, recipes, and customer and supply lists. This information is very valuable to organizations and takes time and effort to develop. Keeping it confidential is critical to your business; otherwise your product or service could lose its competitive edge. Information is not classified as a trade secret if it can be reverse engineered. Reverse engineering is where the competition can, through careful analysis, figure out the unique components, their composition, and relative quantity.

DURATION

For as long as the trade secret remains confidential and is not disclosed or reverse engineered.

◀ **Computer code**
Instead of patents, various aspects of software, including its source code, may be protected by trade secrets. It is generally hard to reverse engineer object code (the sequence of 1s and 0s that execute instructions) to obtain source code. Patents protecting software often disclose algorithms, but this is not the case for trade secrets. Hence, many developers prefer the trade secret route.

Trade secrets refer to confidential business information. This information is very valuable to organizations.

▲ **Secret formulas**

The Coca-Cola formula, Merchandise 7X, is probably one of the most talked-about trade secrets in the world. This secret formula was developed by John Pemberton, a pharmacist from Georgia, in 1886 and has since captured the minds of many people worldwide. The Coca-Cola Company has always remained tight-lipped regarding this formula and it is believed that only a few executives have full knowledge of the secret at any one time.

Information is not classified as a trade secret if it can be reverse-engineered.

▶ **Secret recipes**
Famous recipes such as those for chocolate truffles can be protected by trade secrets. Chef Bill McCarrick, master chocolatier of Sir Hans Sloane Chocolate and Champagne House, says, "My most important new chocolate creations are kept carefully confidential and are protected as trade secrets."

HOW TO PROTECT IT

- Keep it secret.
- Have good practices in place to prove you maintain a secure workplace.

TIPS

- Have business partners, subcontractors, and employees sign confidentiality letters or nondisclosure agreements to show proof of confidentiality.

EXAMPLES

Formulas, the combination of fragrances in a perfume, the ingredients in some chocolate truffles, the know-how that helps you to manufacture a product and get it to market effectively, a proposal for a television series, customer lists belonging to a business, a method of manufacture, marketing plans, survey methods used by professional pollsters.

▲ **Secret recipes**

Colonel Harland Sanders developed the secret KFC formula, which includes 11 herbs and spices, when he operated the Sanders Court & Café from a roadside restaurant in Corbin, Kentucky, in the 1930s. Although this trade secret is now locked up in a vault in Louisville, Kentucky, Colonel Sanders used to carry the secret formula in his head and kept the spice mixtures in his car. Very few people today know the billion-dollar recipe and they have all signed nondisclosure agreements.

► **Secret recipes**

The recipe for Lea & Perrins is a unique trade secret. John Lea and William Perrins, owners of a pharmacy, prepared a sauce according to instructions given to them by Lord Sandys, who had brought the recipe back from Bengal. The sauce, however, was disappointing, so it was left and forgotten in their cellars. Rediscovered after a few years, the sauce was tasted again. Much to the amazement of the pharmacists, the matured sauce had turned into a taste sensation. They capitalized on their find, which became a worldwide success. The recipe has been secret for over 160 years and at any moment only three or four people know what special ingredient gives Lea & Perrins its extra oomph.

Utility patents can cover any invention, but many are in the fields of science and technology.

utility patents

DEFINITION

Utility patents can cover any invention, but many are in the fields of science and technology. They protect a basic product and the way it is made, as well as specific features for that product or process. New patents can cover existing inventions that have been expanded or built on to make them better, easier, cheaper, or more efficient to use. Patents cover hundreds of thousands of products and processes all around you, ranging from vitamin pills to flat-screen televisions, from the windshield wipers on your car to a pacemaker.

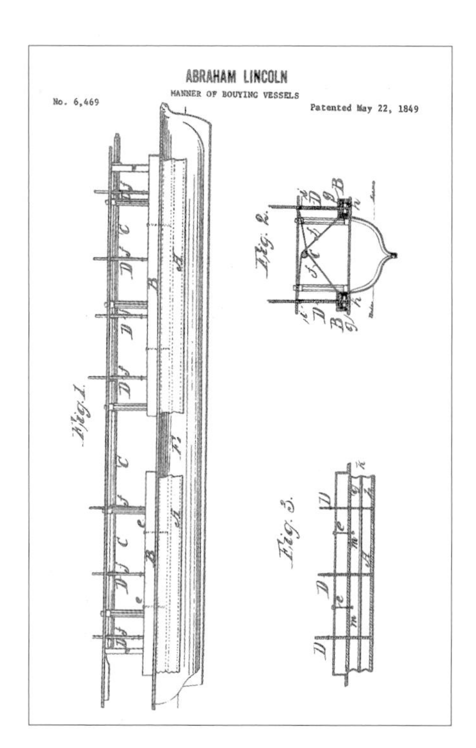

DURATION

Twenty years from the date of filing the patent application at the US Patent and Trademark Office.

◀ **Abraham Lincoln**

The only US president to have held a patent (US Patent No. 6,469), Abraham Lincoln was granted his patent on May 22, 1849 for an invention to lift boats over shoals. A scale model is on display at the Smithsonian Institution in Washington, DC.

► **Jamie Lee Curtis**

Jamie Lee Curtis is the proud owner of US Patent No. 4,753,647, that was granted in 1988 for a diaper with a moisture-proof pocket capable of holding one or more clean-up wipes. Stylish and useful.

◄ **Edison light bulb**

The light bulb is one of Thomas Edison's most famous inventions. A prolific filer of patents, Edison had close to 1,100 patents to his name.

► **VELCRO® fabric**

Inspiration for patent inventions can occur when walking your dog. George de Mestral's dogs would be covered in burs when he took them walking along the foothills of the Swiss Alps. By examining the burs more closely, Mestral noticed that they projected hooklike teeth, which enabled them to stick firmly to the coats of his pets. He was thus inspired to invent a two-pronged fastener. He coined the word **VELCRO** as a trademark; a combination of the words "velour" and "crochet." The patent was granted in Switzerland in 1954.

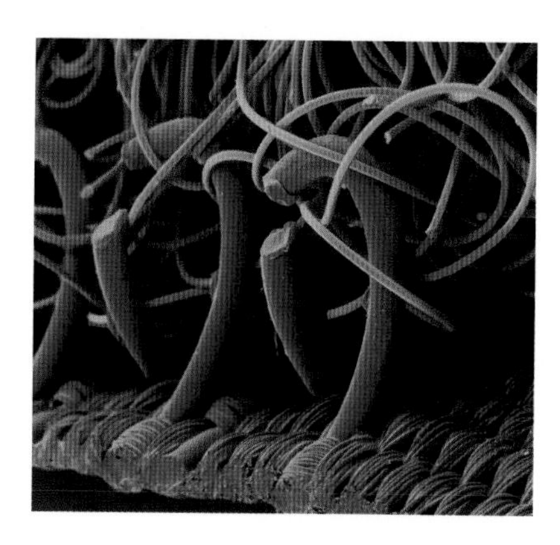

HOW TO PROTECT IT

- File a utility patent application through the US Patent and Trademark Office.
- This must be done within one year of the first offer for sale or public use in the US or publication anywhere in the world.

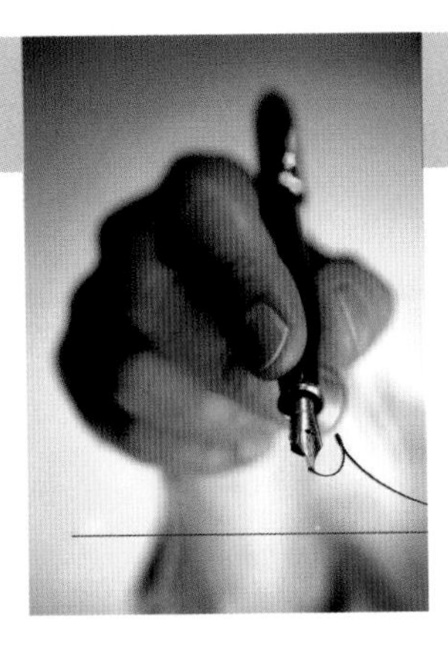

▲ **Fountain pens**
Everyday objects such as pens often embody patented inventions. Famous names like Parker, Sheaffer, and Waterman dot the patent landscape of the late 19th and early 20th centuries with their fountain pen innovations. Laszlo Biro later patented the modern ballpoint pen, known simply as a "Biro" in some countries, before selling his rights to Bic.

TIPS

- Run to the US Patent and Trademark Office to beat the competition.
- Before filing, keep good records and date them.
- Consider a patent search before filing to avoid disappointment or huge expenses.
- Don't disclose your invention before filing!
- Have business partners, subcontractors, and employees sign confidentiality letters or nondisclosure agreements until you are on file.

EXAMPLES

The telephone, the light bulb, online retail procedures, pharmaceuticals, certain computer programs, ballpoint pens, vacuum cleaners, windsurf boards and rigging, the computer mouse, the cut of a diamond, artificial heart valves, methods for downloading music, plastics.

They protect a basic product and the way it is made, as well as specific features for that product or process.

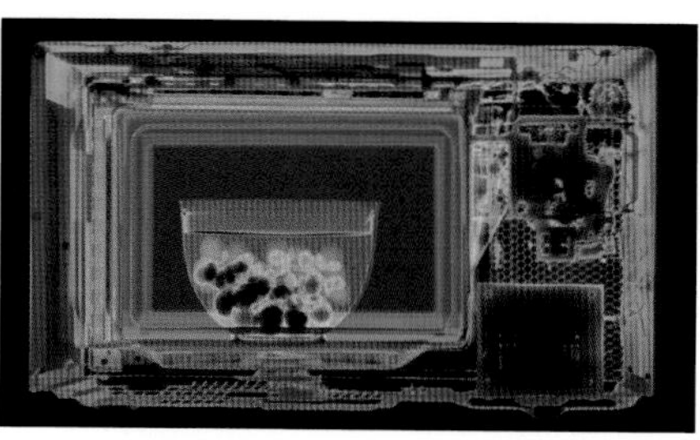

▲ **Microwave oven**
As with many great inventions, the microwave oven was an accident. The inventor, Percy Spencer, was surprised to notice that, while he was standing near a magnetron (which generates microwaves), the chocolate bar in his pocket was beginning to melt. An avid foodie, he next tried popping popcorn kernels the microwave way before rushing to the patent office in Washington, DC, on January 24, 1950 to file a patent application for microwave energy to cook food (US Patent No. 2,495,429).

▲ **Diamonds**
Even diamonds may be the subject of patent protection. New laser technologies that micro-inscribe diamonds by graphitizing the diamond's surface are protected by patents.

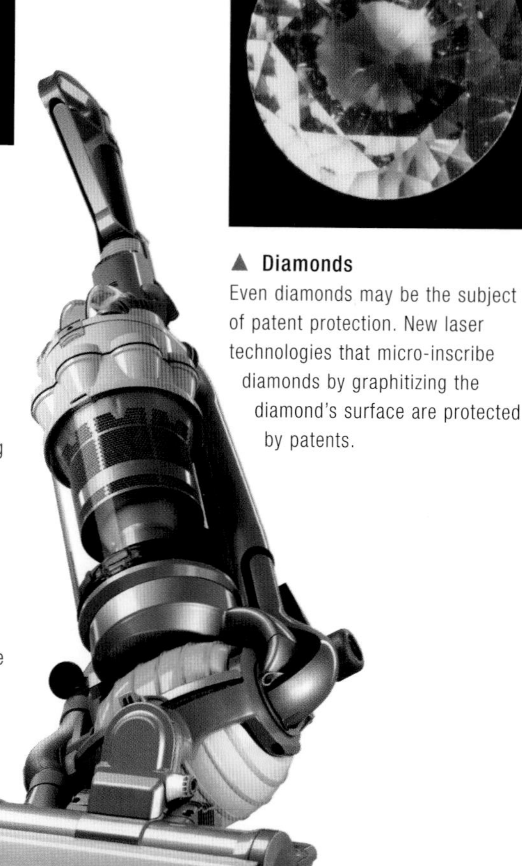

► **Vacuum cleaners**
Dyson's vacuum cleaner revolutionized the vacuum cleaner market by using G-force technology. No dust bag is necessary and all the dirt is collected in the cylindrical body. The centrifugal spin keeps the air stream clear, which means that there is a continuous, strong suction force.

Inevitably, we all want to pigeonhole things. It seems to make life easier. But it can restri and hamper us. With that said, a summary of the categories covered in this book are set out below. Don't study this too hard at this stage—it is a starting point. When you finish the book, look at this chart again. Let the words and concepts flow together to create a whole—you will then be on your way to using intellectual property to your benefit.

summary

TRADEMARKS

Duration	What it covers	How to protect it	Tips
Indefinite as long as the trademark is used	Words, designs, and other markings that identify and distinguish products or services from one another: **HEINZ**, **iPOD**, the **NBC** peacock logo, the **COCA-COLA** bottle, the shape of the **ROLLS-ROYCE** hood, **DHL**, **TINA TURNER**, the color of **VEUVE CLICQUOT** labels, the three rings at the center of a **MONTBLANC** pen	▷ Through use of the mark on your product or service ▷ Seriously consider registration with US Patent and Trademark Office	▷ Select a strong mark ▷ Keep records of use of the mark and especially the date of first use ▷ Obtain a trademark search before use or filing to avoid disappointment and to evaluate risk ▷ Consider adopting the same domain name or trade name as your trademark ▷ Police and enforce your trademark ▷ Use ® for registered and ™ for unregistered trademarks as a warning

RIGHTS OF PUBLICITY

Duration	What it covers	How to protect it	Tips
Varies from state to state. In some states the right dies with you; in others it lasts for 100 years after your death	The name, pictures, voice, or characteristic of any celebrity or someone who has "selling power"	▷ There is no federal system in place for these specific rights, but you can rely on many state laws ▷ Indiana and California have especially helpful laws ▷ California, Texas, Nevada, Indiana, and Oklahoma have name registries to support the rights of the heirs of dead personalities	▷ If you use or plan to use or license your name as a trademark for specific goods and services, file a federal trademark application ▷ Get domain name registrations before you become too popular

COPYRIGHT

Duration	What it covers	How to protect it	Tips
Life of creator plus 70 years. Or, for works created by employees for their employer or works for hire, 95 years from publication	Creative expressions in some type of medium: literature, writings, computer software, visual designs, logos, photographs, website designs, music recordings, motion pictures, jewelry designs, architectural drawings	▷ Automatic on date of creation ▷ Consider obtaining a registration through the US Copyright Office to assist proof of ownership and make automatic damages available if copied	▷ Date and sign or stamp to show proof of ownership ▷ Consider obtaining notary or third-party confirmation of date of creation or an affidavit ▷ Use a copyright legend as warning: © 2007 John Smith. All rights reserved

TRADE SECRETS

Duration	What it covers	How to protect it	Tips
As long as the information remains confidential	Confidential business information: formulas, technical know-how, computer code, recipes, customer and supply lists	▷ Keep it secret ▷ Have good practices in place to prove you maintain a secure workplace	▷ Have business associates, subcontractors, and employees sign confidentiality letters or nondisclosure agreements to show proof of confidentiality

UTILITY PATENTS

Duration	What it covers	How to protect it	Tips
Twenty years from the date of filing the patent application at the US Patent and Trademark Office	Inventions: the telephone, the light bulb, online retail procedures, pharmaceuticals, certain computer programs, **AMAZON**'s **ONE-CLICK CHECKOUT**, **VIAGRA**, the ballpoint pen, the cut of a diamond, artificial heart valves	▷ File a utility patent application with the US Patent and Trademark Office ▷ Must be done within one year of the first offer for sale or public use in the US or publication anywhere in the world	▷ Run to the US Patent and Trademark Office to beat the competition ▷ Consider a patent search before filing to avoid disappointment or huge expenses ▷ Don't disclose your invention before filing! Have everyone sign confidentiality letters or nondisclosure agreements until you are on file

DESIGN PATENTS

Duration	What it covers	How to protect it	Tips
Fourteen years after the design patent issues	Ornamental features on manufactured articles: design of a perfume bottle, ornamental necklace, the shape of a new car, distinctive packaging for a product, the shape of a cookie, the shape of a new handbag	▷ File a design patent application with the US Patent and Trademark Office ▷ Must be done within one year of the first offer for sale and public activities and six months after any foreign filings if they issue during that period	▷ Run to the US Patent and Trademark Office to beat the competition ▷ Consider a patent search before filing to avoid disappointment or huge expenses ▷ Have everyone sign confidentiality letters or nondisclosure agreements until you are on file

bundling your assets

TAKE A HOLISTIC APPROACH

The concept of bundling your intellectual property rights involves looking at your intellectual creation from a holistic perspective. You may have rights to more than one form of intellectual property, and these can be bundled together to help strengthen your arsenal. The more intellectual property rights you have established in your new creation, the better are the chances of an infringer being caught transgressing one of the boundaries. Believe me, in any intellectual property conflict you need strong and varied weapons—whether you are involved in the full-scale war of an infringement suit or the more subtle art of settlement negotiations.

The Panerai-branded watch opposite provides an example of the different forms of intellectual property rights that may be embodied in a single creation (see pp. 236–237 for more examples). You may wish to go through a quick checklist (see p. 46) to see whether your new creation qualifies for protection in more than one category of intellectual property.

Q: How many forms of intellectual property do you think can be embodied in this watch?

A: There can be at least ten!

1
Design patent rights in the watch-face design

2
Trademark rights in the word mark "Panerai"

3
Trade name rights in the company name "Officine Panerai"

4
Trade secret rights in the manner of assembling the watch mechanism

5
Trade secret rights in the manner of manufacturing some of the parts

6
Utility patent rights in the crown lock

7
Design patent rights in the watch-case design

8
Copyright in the design drawings of the watch

9
Copyright in the advertisements that feature the watch

10
Three-dimensional trademark rights in the product design for the watch

USE MULTIPLE LAYERS OF PROTECTION

The bundling of intellectual property rights is not a new idea. In fact, almost a century ago the **COCA-COLA** Company was already using the bundling strategy very effectively. On the advice of its intellectual property lawyer, the **COCA-COLA** Company developed the distinctive "contour **COCA-COLA** bottle" in 1915 in order to distinguish it from imitators and infringers in the marketplace. The **COCA-COLA** Company filed a design patent application, which protected the ornamental shape of the **COCA-COLA** bottle for a period of 14 years. Later on, the **COCA-COLA** Company was also able to register the distinctive "contour" bottle shape as a trademark. By this time, through extensive use, the shape of the bottle had acquired a reputation and identified and distinguished **COCA-COLA** in the minds of the average consumer.

In other words, the Coca-Cola Company succeeded in bundling two different intellectual property rights in the contour bottle. This bundling strategy resulted in the protection of the shape of the bottle well beyond the original 14-year period.

HERE'S COKE . . . THE
PAUSE THAT REFRESHES
Coca-Cola
"Coke"
Ask for it either way . . . both
trade-marks mean the same thing.
5¢
COPYRIGHT 194_, THE COCA-COLA COMPANY

key tips

1 DON'T DILLY-DALLY

All intellectual property rights are time critical, so do not procrastinate. So many intellectual property conflicts and cases boil down to one very simple issue—who was first in time to create something and who was first to file with the appropriate government office to formally protect their rights. Our advice to you is simple: run, don't walk, to the US Patent and Trademark Office and the US Copyright Office and beat your competitors by protecting your new creations formally.

For some categories of intellectual property, you automatically get rights whether or not you get formal protection. This applies to trademarks, where you can develop rights from use, and also to copyright, where your rights exist from the moment of creation. So you might ask, why bother with registering these rights? Getting registrations helps pinpoint when you had what rights and what these rights are at a given moment. And for trademarks, filing an intent-to-use application will get your foot in the door before you have any use. Also remember that a trademark registration gives you nation-wide rights. Even if your rights are prior to those of someone else, it is amazing how much money can be spent proving that you were first. And don't forget that fighting a dispute takes you away from your business. Filing trademark applications for your important brands and copyright applications for your important creative works can save you lots of headaches in the long run. If your creation includes an invention or an ornamental design, and you let one year pass from your disclosures, you lose everything.

And if you are interested in international protection, you have to get on file before any types of disclosures are made, or you risk losing your rights immediately. So, especially with patent protection, it doesn't pay to delay.

2 BITE YOUR TONGUE

If you have come up with a new creation, you are proud, or excited, or enthused, or all of the above. You also want other people to try it out to see how they react or to make sure it works. But, if you are not careful with how you handle your creation before you start filing and taking other precautionary measures, many of your rights could be at risk. As we have said before, this is not the time to show off. Save that for later.

Generally speaking, for virtually all forms of intellectual property, our strong advice to you is to play things close to your chest. Don't allow someone to run away with your ideas. And don't get in a situation where you lose sleep thinking it could happen. This is particularly necessary when you feel compelled to discuss your concept or idea with someone else: we all need sounding boards, and protecting an idea or concept before it has been written down or coded into a tangible medium is often very hard. But be professional: first go through the formalities of identifying and establishing your intellectual property rights, then make sure that the people you need to talk to sign nondisclosure agreements (NDAs) (see Appendix 4, pp. 270–271), and only then can you can start bouncing ideas off them.

3 KEEP GOOD NOTES

Good record-keeping and good housekeeping are essential. Obtaining and enforcing your intellectual property rights will often come down to producing the necessary evidence to demonstrate convincingly that your rights are prior to someone else's. This is where the good note-taker is in the driver's seat. You can almost never go wrong if you keep careful records from the date of inception and creation of your new intellectual property. Keep full and accurate records of all details that relate to the origination and continued use and development of your new creation.

checklist

- ☐ Have you identified all the types of intellectual property that you hold? Are there any areas of crossover, for example, a trade secret that could be patented, or a design patent that could also be trademarked?
- ☐ Do you have protection for your intellectual creations or business assets? Where possible, have you registered your intellectual property?
- ☐ Do your agreements with business associates, subcontractors, and employees clearly define the ownership of the intellectual property at stake?
- ☐ Do you have evidence to prove your ownership of your intellectual property? Are you keeping accurate records to support your evidence of ownership?
- ☐ Have you checked what your competitors are up to? Have you searched public registries and databases to see whether someone has filed or owns conflicting intellectual property rights?
- ☐ Are you making use of someone else's intellectual property in your own creations? If so, do you have the necessary clearance or consent?
- ☐ Are you policing your intellectual property to watch out for someone else infringing your rights?
- ☐ Have you used the appropriate warnings or notices to alert everyone to the intellectual property rights in your creations?
- ☐ Are you administering your intellectual property to ensure that the necessary rights are maintained and renewed in good time?
- ☐ Do you have the necessary safeguards in place to keep your creation confidential?
- ☐ Do you have good legal counsel to guide you through all the above?

CHAPTER 1
trademarks

WHAT IS A TRADEMARK?

The alarm on your **RADIO SHACK** radio goes off. Your local **NPR** station comes on and you listen for a few minutes to **"MORNING EDITION."** You go into the kitchen, get some **STARBUCKS** ground coffee out of your **GE** refrigerator and start brewing it in your **BRAUN** coffeemaker with a **MELITTA** filter. While that's brewing you flip on your **SONY**

You have been up for 10 minutes and have already encountered scores of trademarks—words, slogans, logos, stylized letters, colors, package designs, and even product designs that brand owners have developed and use to sell their products and services to you. You as a consumer have a choice as to which brands you want to rely on. What drove you to these—images, reputations, little choice, happenstance, or habit? If you have any type of product or service that you want to sell, it pays to understand how to select, use, protect, and enforce your trademarks. The owners of the above trademarks have taken a lot of care and time doing this. So should you.

television and check out **THE WEATHER CHANNEL**. Next, you cut a slice of fresh **ENTENMANN'S** blueberry coffee cake you got last night at **KROGER** with your **HENCKELS** knife and place it on your **MIKASA** plate (you are off the **ATKINS** diet this week). You sit down in your **IKEA** chair and start to review your email messages on your **HP** laptop. **YOU'VE GOT MAIL**.

A trademark is a word (like **MCDONALD'S**) or design (like the **NIKE** Swoosh) or other indicator (like the **TARZAN** yell or the **TIFFANY** blue box design) used with particular products or services. A trademark is the link between the manufacturer or service provider and the customer. It helps customers distinguish competing products and services from one another (like **COKE** versus **PEPSI** or **VISA** versus **MASTERCARD**). Trademarks can also suggest a certain cachet (**NEIMAN MARCUS** vs. **WAL-MART**). At the same time, trademarks can be used for any type of product or service, not just luxury or expensive items—think of **BIC**, the workhorse of pens, versus **MONTBLANC** crafted

A patent covering a particular product may have expired long ago. But the trademark for it lives on and can drive customers to that product instead of competing products.

writing instruments. Once a trademark develops a reputation, it will also give customers an idea of what to expect when they see the trademark on new products and services. Think of **BIC** disposable razors and **MONTBLANC** watches. It is more than the words that let you know what brands these are. The logos, type styles, and other symbols all reinforce your comfort level with these products.

Once a trademark develops a reputation, it will also give the customer an idea of what to expect when they see the trademark on new products and services.

Trademarks also serve an important public interest: to help prevent consumers from being deceived or misled into buying counterfeit products or services. When you buy a bottle of **BAYER** aspirin for a headache, you expect it to work the same way it has for you in the past. But there are actually counterfeit drugs out there, and you would be very unhappy if you took a fake **BAYER** aspirin that gave you liver disease.

Unlike virtually all other intellectual property rights, trademark rights last for as long as the trademark is used—potentially forever! The federal registration for **ZILDJIAN** cymbals alleges use back to 1603. Further, trademarks are protected as soon as you use them. They do not even need to be registered. Use is the most critical step in developing trademark rights. You need to use a mark to protect it. And you need to keep using it to maintain these rights.

A patent covering a particular product may have expired long ago. But the trademark for it lives on and can drive customers to that product instead of competing products. Many people would prefer to start off the day with **KELLOGG'S** corn flakes, rather than some other brand, even though the corn flakes patent expired long ago. No wonder a trademark can be considered a more valuable asset than a patent.

How are trademarks used?

Trademarks do not need to be used in connection with just a single product or service. They can cut across entire groups or categories of products (**JOHNSON & JOHNSON**) or, conversely, they can be very specific, like a model name (**CAMRY**). Even ingredients of products can be trademarks (**SPLENDA** sweetener in a diet beverage). And what about business names and domain names? If they are distinctive enough and used in a certain way, they too can be registered as trademarks. Even if they cannot be registered as trademarks (and this is discussed further below), the law can still protect them. Keep this in mind when you are selecting your brand. Do you want the brand to also be the name of the company (like **KODAK**) or a house mark for everything (like **GENERAL FOODS**)? Is it just a specific product name like **GRAPE-NUTS** for a cereal? Will it be a style name like the **F-100** of Ford Motor

In this crowded marketplace, anything you can do to separate and highlight your product or services from those of your competitors will pay off.

Company, an ingredient for other people's products like **NUTRASWEET**, or will it be a domain name like **AMAZON.COM**? You may not know when you select the brand whether or not it is going to cross over and be used in these various manners. But it is not a bad idea to give this some thought.

Not any old word, design, or symbol is automatically a trademark

It is easy to talk about trademarks that we all know and recognize. But when a business is getting started or a particular product line is being launched, an important selection process occurs. Trademarks don't grow in pumpkin patches! Instead, if it is done right, the trademark will link the manufacturer or service provider to the customer base and help customers come back to that brand over and over again. This is a lot easier said than done. For example, will a customer recognize your new trademark as referring to your particular product or service alone? Will your trademark enable customers to form an exclusive association with your goods or services? If the answer is yes, your mark is probably distinctive and you will be on your way to building a valuable asset. Some brand owners go for descriptive marks because they immediately inform the customer about the product. Sometimes that works—you need to know your market and your customer base. You should also know that descriptive marks are harder to search, register, and protect, and cost more to enforce.

But any old word, design, or symbol can become a strong trademark if it is distinctive

You should not underestimate the types of trademarks you can use to build your brand. There has been a revolution in types of marks over the years. Do not underestimate how logos, slogans, color, packaging design, product design, sound, and even smell can be used alone or in combination to create a strong brand. These can be trademarks if the owners put enough money and advertising behind them. Just look at the color brown campaign of **UPS**. In this crowded marketplace, anything you can do to separate and highlight your product or services from those of your competitors will pay off.

Will your trademark enable customers to form an exclusive association with your goods or services? If the answer is yes, your mark is probably distinctive and you will be on your way to building a valuable asset.

Examples of trademarks

Word marks like **HEINZ**, logos like the **CADILLAC** medallion, sound marks like the ding-dong-ding chimes of **NBC**, packaging marks like the **COCA-COLA** bottle, color marks like brown for **UPS**, celebrity names like **TINA TURNER** and **MICHAEL DOUGLAS**, character names like **MICKEY MOUSE**, and domain names like **AMAZON.COM** are all trademarks.

CATEGORIES OF MARKS

Armed with this arsenal of potential trademarks to create your brand or revitalize or update your existing brand, the selection process goes into high gear. What type of word will best help you sell your product or service? Is there an icon or logo to associate with your brand without even seeing the word mark (think of **MCDONALD'S** Golden Arches)? If you are a retailer or have an important product to sell, do you want your packaging alone to help sell the product without the consumer even seeing the label or name of the product (think of the **COCA-COLA** bottle)? At this stage of the game, the choices seem endless. But all trademarks, unfortunately, are not created equal. Set forth below are the general categories of marks that the trademark world has come to rank in a pecking order.

1. Generic terms

These are terms that refer to a category or type of good or service. For example, you obviously can't trademark the words "software," "jewelry," "furniture," "airline," or "restaurant" for their respective categories of goods or services. Generic terms directly define the product or service; competitors need to use them and consumers will never care about them. Generic words as trademarks are a big no-no.

This doesn't mean you can't add a generic word to a distinctive word to create an overall trademark, as in **SONY PICTURES**. Sometimes a generic word added to a distinctive word can help define your brand in the marketplace.

NOTE OF CAUTION!—Victims of their own success

You always need to be careful about the way you use your trademarks. Some famous trademarks have become victims of their own success. For example, **TRAMPOLINE** was a trademark at one time. **CELLOPHANE** was another one. Trademarks can become generic if their owners are not able to stop customers from using the trademark to refer to the product. This especially happens when a trademark is coined for a brand new product that is patented and there is no competition in the marketplace. To avoid this, make certain that you have a good word to describe your product and keep it separate from the trademark for that product. For example, you do not make a **XEROX**, you make a **XEROX** photocopy.

2. Descriptive marks

Terms that describe a feature or specific quality of the goods or services are difficult trademarks to protect. You can also have a hard time protecting trademarks that are just a surname, describe the location where your

If a descriptive mark is used long enough and you manage to keep others from using the term as a trademark, it can become protectable.

product is made, or terms that sing the praises of your product. These types of trademarks are not automatically protectable because they do not help your customer distinguish your product from your competitors' products. After all, they may need to use the same words to describe their own products. If a descriptive mark is used long enough and you manage to keep others from using the term as a trademark, it can become protectable.

Examples of formerly descriptive trademarks

Here are some good, strong trademarks we all know, but were not born that way—**SOFTSOAP** for liquid soap (descriptive), **THE GREATEST SHOW ON EARTH** for circus performances (sings praises), **POLAND SPRING** for drinking water (geographic), and **MCDONALD'S** for fast food restaurant services (surname). Don't underestimate the amount of time and effort these owners have taken to protect these as assets.

3. Suggestive marks

These are terms that suggest an attribute or benefit of the goods or services without directly describing them. There are not many places in the law where a line is drawn, but in the trademark world a descriptive mark is not automatically protectable and a suggestive mark is automatically protectable.

However, figuring out which side of the line a particular trademark falls is not always easy. There are many examples of suggestive trademarks, including **JAGUAR** for cars, which suggests sleek, fast, elegant; **DOVE** for soap,

Letter combination marks or acronyms can be good trademarks if the acronym itself has nothing to do with the product.

evoking something soft and gentle; and **COPPERTONE** for suntan lotion, which suggests the desired color resulting from use of the product.

Other marks that fall into this category include names and letter combinations. Even though a surname is not automatically protectable as a trademark, an entire name or a first name can definitely be protected right off the bat. For example, **RALPH LAUREN** was a protectable trademark from the beginning.

Letter combination marks or acronyms can be good trademarks if the acronym itself has nothing to do with the product. For example, don't bother trying to use **TV** as a trademark for televisions. But **HSBC** and **YMCA** were protectable marks from the beginning even if Hong Kong Shanghai Banking Corporation and Young Men's Christian Association would have been viewed as descriptive and not automatically protectable.

4. Arbitrary marks

This category is the source of many strong and protectable trademarks. These are dictionary words, terms, or phrases that are familiar to people but are used for goods or services that have absolutely nothing to do with the product or service. These include **SMASHING PUMPKINS** for the name of a music group and **DELTA** for airline services. These types of marks are automatically protectable. You will also have a strong mark that you can enforce against others.

Other marks in this category include made-up names. It is not uncommon for brand owners to adopt fictional names. For example, **BETTY CROCKER** was a fictitious character name created by General Mills but it has become a strong trademark. **WENDY'S** is derived from the name of the original owner's daughter. These are entirely good trademarks. Just make certain that you made it up and that it does not belong to someone else who is already using it as a trademark in your field or in a related field.

5. Invented and fanciful marks/coined terms

These are words that are typically invented or entirely made up. Sometimes they are derived from a combination of bits and pieces of descriptive words for the product or service. Sometimes a computer spits them out. Sometimes they are words spelled backward. Other times they are something that pops out of the head of a marketing genius. Examples include **ROLEX** for watches and **KODAK** for cameras. In the beginning, it takes some strong advertising effort to educate the consumer to associate such trademarks with the products and services. But a good marketing campaign and, of course, a good product or service will get you a great and unique brand. Think of **GOOGLE** for search engines.

It is not uncommon for brand owners to adopt fictional names.

SELECTING YOUR TRADEMARK

So which category of mark is best?

The answer is any category, except of course generic words. Think of **DELTA AIRLINES** versus **AMERICAN AIRLINES**. **DELTA** is an arbitrary mark. From the first use, it would have been easier to protect than **AMERICAN**, which is geographic. But after years of use and promotion, they are now both strong marks.

With that said, descriptive marks are probably the most common category of trademark you will want to select. It is probably because you can immediately tell the customer about a particular feature or quality of your product, where it is made, or the person who designed it.

This is one area where you and your trademark lawyer can get into a tug-of-war. You get one end of the rope. You want a descriptive or suggestive mark because your customer will understand your product just by hearing the trademark. At the other end of the rope is your trademark lawyer. He or she will urge you to adopt a legally strong mark. It will be easier to register and stop your competitors from using. Your lawyer will usually recommend the use of invented and fanciful terms and possibly suggestive words. Who wins? In our experience, the answer often lies in the middle: suggestive marks get out the message, but can be protected trademarks from the beginning.

Whatever words, logos, slogans, and designs you select, remember that it takes a serious commitment of resources and money to create brand awareness to educate the consumer.

However, each commercial opportunity varies. Your product might compete in a market where the profit margin is low, and so is your budget. In these cases, you don't have the money or time to spend building a unique brand. You just need to make sales. On the flip side, you might be launching a new and unique item that will revolutionize the market. Or your company might be headed for an IPO. Here, you may have no choice but to adopt an invented or fanciful mark so that you can distance your brand from copycats and you can impress your investors with a strong potential brand.

When selecting your mark, choose wisely, taking into consideration all the pros and cons mentioned above. Your trademark lawyer knows descriptive marks are tough to protect. These marks need to earn their keep by being used carefully and promoted extensively. At the end of the day, it is your decision. You are the most familiar with your product and how you want to position it in the marketplace. Just because a brand might be difficult to protect does not mean it is a bad one.

Whatever words, logos, slogans, and designs you select, remember that it takes a serious commitment of resources and money to create brand awareness to educate the consumer. And, of course, the underlying product or service had better withstand the competition or your brand will mean nothing to anyone, no matter how much time and effort you devote to brand selection.

DEBUNKING THE MYTH—It doesn't pay to get cute

Some brand owners love to play around with words to spell them differently or pick and choose elements or words to create a distinctive word. For example, **FONE** for a telephone service, **TOYSDIRECT** for a discount toy store, **CAR-FRESHENER** for an odor tree, or **BOOKS.COM** for an online book retailer are not valuable assets and never will be. Sure, a few consumers might start to associate such terms with your goods or services but the Patent and Trademark Office will not register them and most judges are not going to be fooled into thinking that you have something that is protectable. Further, if you ever try to sell your business, don't count on someone paying a lot of money for such a trademark. You'd better have some stronger assets than that!

Other words and symbols that are not protectable

Words, phrases, and other symbols that are immoral, scandalous, or deceptive will not be protected as trademarks. Also, forget about state and national flags and other national symbols. In this day and age, when everyone is trying to stand out from the crowd, some people are tempted to adopt marks in this category. If you do that, you are wasting your time trying to register or enforce your trademark. If you have a serious product or service you want to sell, please stay away from these types of marks.

If you are launching a new product or service or you want to rebrand your existing product line, think of using lettering, logos, slogans, distinctive colors, package design, and even product design.

Don't use other people's names

Don't even think of using names, pictures, or other characteristics of celebrities or other living individuals as your mark, unless you are that celebrity or have their written consent. Your own personal name can be a very good trademark. So can a made-up name. Ripping off someone else's name or image, like a fake Woody Allen voice to sell cars, is not going to get you anywhere except into a lot of trouble.

Titles

Generally, titles of books, songs, albums, or movies are not protectable under the trademark law. But if you use them in a series, like **ROCKY I**, **ROCKY II**, and **ROCKY III**, *voilà!* You have trademark rights. Or if you sell or give away merchandise under the title, you might be able to protect it, at least for the merchandise. Many Broadway show titles can pull this off, like **PHANTOM OF THE OPERA** or **RENT**.

Don't use famous marks for other products and services

The owner of a famous trademark can stop you from using that famous trademark, even if it is for different products and services. For example, you can't use **KODAK** for socks since **KODAK** is a famous trademark. This is known as "dilution." So don't bother trying to use a famous trademark for your own products and services and then argue that your products and services are different from those of the famous trademark owner.

BUILDING YOUR BRAND

Once you have selected your mark, you need to develop it into a brand. At this stage of the game, you should give serious thought to using a graphic designer to create a unique and distinctive presentation for your new mark. Look at the silver foil with blue lettering on a **YORK** peppermint patty or the portrait of a Roman centurion on your **AMERICAN EXPRESS** card. Trademarks are not presented in black typed letters on a white package with no background.

Instead there is a whole supporting cast surrounding the word mark. If you are launching a new product or service or you want to rebrand your existing product line, think of using lettering, logos, slogans, distinctive colors, package design, and even product design.

All of these elements can be combined, and used consistently, to give your product and service a memorable brand identity. They are also trademarks that you can protect and own as assets. If you do it right, some of these supporting cast members will become stars on their own.

Trademarks are not presented in black typed letters on a white package with no background.

TIP

THINK ABOUT COPYRIGHT

Keep in mind that your design marks might also be protected under the copyright law. This gives you another layer of protection for your brand, and helps bundle up your rights. And if your designer is not your employee, make sure to get a written assignment of any copyright in the designs they create for your brand.

Trademarks like color and package design are latecomers to the trademark world. They are important (and welcome) options in this crowded global marketplace where your brand needs to be separated from the rest. Generally, these types of additional marks are considered descriptive. So be patient—it takes time before all elements of your brand will be protected as trademarks. And sometimes you can't even register them right away. But the investment pays off. Think of the blue box with the white ribbon from **TIFFANY & CO**. Here is some more information on the types of elements that can become trademark stars.

▶ Slogans

Think of **DON'T LEAVE HOME WITHOUT IT**. You know it stands for **AMERICAN EXPRESS** even without seeing that mark. It's a great slogan. And it also says something about the service—it's indispensable, it protects you. You can combine your own group of words to create your own slogans.

Some slogans tend to be very descriptive, for example **SERVING YOUR COMMUNITY SINCE 1922**. This isn't the greatest slogan since sliced bread, but it does give the consumer comfort that they aren't dealing with a fly-by-night outfit. And there is also a lot of pride in that message. Still, from a trademark standpoint, it is not distinctive and you would probably need to stay in business until the year 2322 to protect it. If you are going to select a slogan, we think it's a good idea to spend time thinking about what you want to tell your customer about your brand every time they see it. If you

are creative, you can also come up with a message that is so distinctive that customers will automatically think of your product when they see it. Such a slogan will be a protectable asset.

▶ Logos

Logos can be designs without words, word marks presented with stylized letters, or a combination of these. Some logos are designed so that the word portion and design portion can be used together and also separately. This is often done by stacking word marks and designs. Or you can intertwine your word mark with a design. Some design marks become so well known they are icons and you do not even need the word mark. Think of the bullseye logo of **TARGET** or the peacock tail design of **NBC**.

▶ Packaging design

Product packaging, such as the **COCA-COLA** bottle, the **TIFFANY** blue box, and the **CHLOE** perfume bottle are all considered trademarks. Even services can be packaged. Think of theme restaurants, like **PLANET HOLLYWOOD** and other chains, like **IKEA** stores. Package design like this is known as "trade dress." Generally, if you want to develop rights in packaging, the packaging needs to be unusual and it needs to be promoted as your trademark for a long time. Packaging can also be the subject of design patents. This way, you can get design patent protection while you are building up a reputation as a trademark. By the time your design patent expires, if you play your cards right, you will have a protectable trademark in the packaging design.

Single colors can be difficult to protect as a trademark for particular goods and services.

▶ Product design

The shape and design of product can also be protected as a trademark. But this is a lot more difficult than package design. You need to prove that you have promoted your product design as a mark exclusively and for a long time. And if your product design has functional aspects to it, or if your consumers just don't think of the product itself as yours, then forget about ever developing trademark rights.

Whatever you do, if you come up with a good product design, and it is unique and not obvious over other product designs already out there, you should try to get a design patent. If the product has some functional improvements or novel features, you may even be able to get a utility patent (getting a utility patent on a product, however, will almost always kill your chances of getting a trademark). When you start to get overlapping protection, building up your arsenal, you are bundling your rights. These will help build assets for your business. It also sets you apart even more from your competitors. Whenever you are in the development stage with a new product and its branding, take a holistic approach. Always look at everything you are doing and then think about the types of opportunities you have to build up a whole portfolio of intellectual property.

Sometimes it is not so clear whether you have developed a product design or a package design. This can be an important distinction, since product design is much more difficult to protect as a trademark than package design.

> **TIP**
>
> **MAINTAIN CONSISTENCY**
>
> Once you have all these elements together and have a brand identity, don't start changing everything. Consumers like consistency. The only reason that **KODAK** has rights in yellow and red for its packaging is because it does not use blue and purple for one type of product and gray and green for another. The **COCA-COLA** bottle is a trademark because **COCA-COLA** makes sure you see it all the time on bottles. It even uses it as a logo on its cans!

One way to answer the question is to stop and think about the goods for which the design is used. If the goods are the product itself, you have a product design. For example, the **COCA-COLA** bottle is not the product being sold. It is just packaging for the carbonated beverage. But if the **COCA-COLA** bottle was the standard-shape bottle for all companies' carbonated beverage brands, the bottle would be a normal part of the product and it would be almost impossible to protect—think of the generic milk carton. No one has trademark rights in that.

► Color marks

Single colors can now be protected as trademarks in the United States (for example, **VEUVE CLICQUOT** has registered its orangey-yellow color for champagne). Single colors can be difficult to protect as a trademark for particular goods and services, partly because there are so few colors to go around! So you can use shades of color. Better yet, think of using color combinations. These are more unique. We believe color combinations make a lot of sense to develop as a trademark, as it increases the potential number of "color" marks available to brand owners and for consumers to get to know. For example, green and yellow are associated with the **BP** brand, yellow and red are associated with the **KODAK** brand. If you decide to add a color or combination of colors to your branding effort, remember you will need a good ad campaign to teach the customer that the colors are associated with your goods and services. Don't count on customers figuring it out on their own.

▸ Sound marks

These are not very common, but you know them when you hear them. Think of the **TARZAN** yell, or the **NBC** chimes. Of course, to develop these sorts of rights, you need radio and television spots that can be heard or a product that makes the sound.

▸ Smell marks

A few people have used a smell as a trademark for products that usually do not have a smell. But don't get the idea that if your product is a perfume you can get trademark rights for your particular scent. The law has not gone that far yet!

▸ Moving image marks

We are getting into some pretty unique territory here. But some people have managed to develop a brief series of movements or images that together make up a trademark. If you have ever seen a motion picture produced by Jerry Bruckheimer, you will know what we mean. Go to www.jbfilms.com. Now that's a trademark!

SEARCHING

Now that you have selected your word mark, symbols, slogans, colors, logos, packaging, and other elements that are going to identify your brand, it would be a good idea to do some searching. Searching is not a requirement before filing an application or before using your marks. But if you launch your brand and someone else already uses or has registered the same or a similar trademark for the same or similar goods, you could have a serious problem. That person might try to stop your use and demand a lot of money for damages. He or she can also oppose the registration of your trademark. So it pays to do some searching before you launch your brand or file an application. It also pays to use a lawyer who knows about searching. This is serious business and is a step you should not ignore.

Preliminary searches

The purpose of a preliminary search is to give you or your lawyer an awareness of similar marks that are already registered. And with the Internet, you get a good glimpse of unregistered marks. The Patent and Trademark Office has a very good search engine for word marks at www.uspto.gov (scroll down to "Trademarks" and click on "Search trademarks"). Professional search firms also have databases and Internet search engines, like **GOOGLE**, **YAHOO!**, or **MSN**, can also help you determine whether an unregistered mark in in use. **DOGPILE** is also useful for unregistered marks. If you know of a good online retailer for your product or services, use that too.

Even if you find a registration of the exact mark for the same goods or services, it does not always mean that you are out of luck. Many trademark owners discontinue the use of their registered marks and these registrations can be subject to cancellation. Or the registration might have recently expired. The Patent and Trademark Office records are usually behind a

If your trademark survives the preliminary search, it is a good idea to get a full search, known as a "comprehensive search."

number of months. You can also use a private investigator to get information on a mark that comes up in your searches to find out whether it is really in use.

Full searches

If your trademark survives the preliminary search, it is a good idea to get a full search, known as a "comprehensive search." Do this with your word marks, your slogans, and even any acronyms you think you might want to use. This can be very useful and can let you know about potential problems before you invest a lot of money in your brand. These are available from several established search firms, such as Thomson CompuMark (www.thomson.com), Name Protect (www.nameprotect.com), and CT Corsearch (www.ctcorsearch.com). They are not lawyers. Instead, they provide a report, often 100–300 pages long, with numerous references to registered and unregistered marks and names from various sources. When you order the search, be sure to tell the searcher or your lawyer the real products and services you plan to market. Otherwise, you will get back an encyclopedia of irrelevant marks. Remember: garbage in, garbage out.

When the report comes in, you and your lawyer should review it carefully. This is a critical stage of the process and you need to pay close attention to the results of the search. And you should understand your lawyer's advice. Don't assume that someone whose mark comes up in the search will never find out about you. Or don't assume that someone is too small a user to give

TIP

FEDERAL TRADEMARKS

As much as we love a federal trademark registration, we don't want to give you the idea it is the be-all and end-all. Registrations are not a guarantee that you can use your marks trouble-free. Prior users can try to cancel them. So can your competitors if they think your mark is descriptive. But your registration will give you the leading edge to stop others from using the same or similar marks for similar products. And, like scotch, the older it gets, the better it is.

you a hard time. Adding other words to your mark or adding logos or other designs might help you differentiate your mark. But this requires the help of a trademark professional. It is not wise to do this part on your own.

You can also get searches for logos and designs. As a practical matter, logos and designs are harder to search. These searches are generally limited to federal registrations and pending federal applications.

NOTE OF CAUTION!—There are no guarantees

Unfortunately, not even a full search is a guarantee and you can never know for certain that you will not have problems with your mark. If your mark has survived the full search, but you have a funny feeling that someone else might be out there, place an announcement in a trade publication or make a press release. This sometimes draws out owners of marks who did not come up in the search. Before you do that, make sure to be on file with your application and that you have any domain names you want. Or a pirate might do this for you!

HAVE A LITTLE PATIENCE

Don't expect to call your lawyer and just tell him or her to register your trademark. Your lawyer will need all sorts of information. So, have a little patience. This isn't like ordering a hamburger and milkshake for lunch.

HOW DO I PROTECT MY TRADEMARKS?

Unregistered trademark rights

The United States, being what it is, is pretty different from the rest of the world when it comes to protecting trademarks. In many countries you need to register your trademarks through the government before you can stop others from ripping you off. In the United States, it is not a requirement.

DEFINITION—Common law rights

If you use your trademark for your products and services, you have rights in your unregistered trademark. These are known as common law rights. These rights can be very strong. But if you do not plan to register your trademarks, keep good records of how you have used them, what products and services you have used them on, how many people have bought these products and services, and the geographic locations where you have sold and advertised your products and services. Otherwise, if you ever want to rely on these rights against someone who you think is violating your rights, you are going to have lots of problems proving your common law rights.

Federal trademark registrations

There are hundreds of thousands of unregistered trademarks out there. And all those owners have common law rights—whether they know it or not. There is nothing wrong with that. But if you are trying to build a good business, or if you plan to sell nationwide, or if you want competitors to stay away from your mark, relying on your common law rights has a lot of

BE SPECIFIC

It is best to give an exact date: month, day, and year. If you are not positive about your use dates, you can just give a year, like 1998. But when you do that, the government treats it like December 31, 1998. You can also use a month and a year, like March 1970, but then the government will treat it as March 31, 1970.

downsides. Getting a federal trademark registration through the US Patent and Trademark Office (a federal government agency located in Alexandria, Virginia) gives you rights across the entire United States. This includes Puerto Rico and other US territories. It is also an asset that you can sell with your business. And it lets other people know about your rights. And if you need to sue someone, waving your federal registration in their face, or showing it to the judge, usually gets you a lot of mileage fast.

And it's not just your word marks that you can register. The Patent and Trademark Office will let you register logos, package designs, colors, moving images, smells, sounds, and sometimes even product designs. Visit www.fromedisontoipod.com for an example.

GETTING A FEDERAL REGISTRATION

It makes a lot of sense to use a trademark lawyer to assist you in filing your trademark application. There are many areas where you can go wrong here. The government will be quick to take away your registration or application if you give it false, or even incorrect, information. You can file trademark applications electronically by going to www.uspto.gov, you can file in person, or you can mail them in. Electronic filing is the way to go—and you get your serial number right away.

Whether you do this on your own, or through a lawyer, here are some decisions you will need to make when you file a trademark application.

Getting on file

Your basis for filing The Patent and Trademark Office needs to know why you deserve to get a trademark registration.

Use You can get a federal trademark registration by proving you are using your mark for your products and services. In order for the federal government to be involved in an application, you need to ship your products within the US or into or out of the US from or to a foreign country. This is known as use "in commerce." You also need to give the government the dates you first used your mark anywhere in the world, the date you first used it "in commerce," and a sample of your use. The Patent and Trademark Office likes digital photos that show your mark on a label or package for your product. They will also accept an ad for your service, including an ad on your own website. Other types of samples work too, but these are the least likely to prompt questions.

Intent to use If you haven't started using your trademark yet, but know of a specific trademark you want to use and register, you can file your application under the "intent-to-use" laws. This is a great opportunity if you are getting ready to do business. To get your registration through, eventually you still need to prove use. There's no free lunch in the Patent and Trademark Office.

Foreigners get all the luck If you are based outside the United States, you can actually get a federal trademark registration if you own a registration in

DRAWING

A typed drawing is the best way to file for word marks. It will cover different visual presentations of your mark. It doesn't mean that you need to present it on packaging and ads as typed letters.

the country where you're based and if your country has a special agreement with the United States. But don't be fooled—you'd better meet all the other requirements. And be prepared to use your trademark soon after you get your registration, or people can come after you and cancel it. If you do not use a mark for three years, it is presumed to be abandoned.

Applicant The "applicant" on your application is the owner of the trademark. This can be an individual, corporation, LLC, partnership, or other legal entity. The owner controls the use of a mark and is responsible for the quality of the goods or services the mark is used with. This is not always easy to figure out. If you are a sole proprietor, or a celebrity, or an individual using a trademark, you can file the application in your personal name. However, in this day and age of product liability lawsuits and tax and estate issues, many people choose to file in a corporate name. You should go over this with your business lawyer or your accountant, since, once you file, it costs money to change the owner. And if it ends up that you file in the name of the wrong owner, forget about it. You have to start all over again.

NOTE OF CAUTION!—Use it or lose it

For all practical purposes, intent-to-use applications cannot be transferred or sold to anyone until after you have proved to the Patent and Trademark Office that you are using your mark.

Goods and services and classification

Remember, in trademark land you do not own a word. Your trademark rights are for the products and services you use or plan to use your trademark on. So the government needs a good list from you. It refers to your "products" as "goods" when you file. You need to be pretty specific. You also need to tell the Patent and Trademark Office which numerical class or classes your products and services fall into. All the products and services in the world fall into 45 classes (see Appendix 1, pp.258–263). Most countries of the world are pretty much in agreement on this. It is probably one of the few times humankind has come together! Your lawyer or the Patent and Trademark Office website (go to http://www.uspto.gov, scroll down to "Trademarks" and click on "Indentification Manual") can tell you which class your products and services fall into. Don't plan to fight anybody about this—they have their minds made up.

Drawing You need to file a page showing the mark—called a "drawing"—with your trademark application. You can file a typed page or a logo page. If you type your mark, the registration will cover the different stylized ways you present it. If you want to protect your mark in a special typeface, design, or logo, you should put that on the drawing page.

Filing fee To file, you have to pay the Patent and Trademark Office a fee for each class. It is in the $300–$400 range per class. You can file one application in several classes if you want. Or you can file one application for each class.

Remember, in trademark land you do not own a word. Your trademark rights are for the products and services you use or plan to use your trademark on. So the government needs a good list from you... You need to tell the Patent and Trademark Office which numerical class or classes your products and services fall into. All the products and services in the world fall into 45 classes. Most countries of the world are pretty much in agreement on this. It is probably one of the few times humankind has come together!

OFFICE ACTIONS

If it ends up that your trademark is descriptive, you might be forced to register your mark on the Supplemental Register. This is a sort of purgatory for trademarks where your rights are pretty limited. But it is better than nothing and you get to use the ® mark. If you use your trademark for a long time—about five years—you can try to register your mark again on the Principal Register. This is nirvana for trademarks.

Declaration You need to sign a declaration that says everything you are saying is true. So no lying or fibbing when you file. If you are in a rush to get on file and do not know your basis, you are allowed to file if you have the mark, the goods and services, and the name of the applicant. You can provide the basis later. But we think it is better to have your act together when you file. It helps avoid fatal mistakes.

Examination process

Within a few days of filing your application, the information you supplied becomes available to the public. Go to http://tarr.uspto.gov/, type in your application number and you can see what's going on with your application.

You should check the status of your application every six months until the registration is issued. You might be thinking, every six months? Doesn't the registration come through in six days? Unfortunately not. Currently, the Patent and Trademark Office is taking about six to eight months to examine applications. They swing back and forth on the time to examination, depending on how much is going on. After that, there is still more red tape. Here's a summary of what happens.

Office Actions

Once your application comes up for examination, an Examining Attorney will review it. If there is a problem, he or she will call or, more likely, send you or your attorney a letter or email called an "Office Action."

TIP

EXAMINING ATTORNEYS

Remember, Examining Attorneys are usually nice people and want to help you. But at the end of the day, they have a job to do, and if you mess up, you might have to start all over again. Whatever you do, even if you think the point an Examining Attorney is making is wrong, do not start to fight. It's OK to disagree. It's America. But keep your cool. It does not make any sense to fight and does not get you any further in the long run.

These usually focus on the following:

- Federal applications and registrations of the same or a similar mark for the same or related goods on file before you. If you do your homework and do a preliminary search before you file, you will likely survive this step.
- The way you word your goods and services. They also make sure you placed them in the correct class.
- The category of the mark. Here's where selecting a mark in the suggestive or fanciful categories pays off. If your mark is descriptive of your goods, or is a surname, or sings praises too much, or describes where the product is made, the Examining Attorney might stick a big red **STOP** sign in your face.
- If your mark contains a descriptive or generic word like "**RECORDS**" for musical sound recordings, you will be asked not to claim exclusive rights in that word.
- Your samples of use. The Examining Attorney will question samples that are just drawings or mockups or anything else that doesn't look like real use.

Responses You generally have six months to respond to Office Actions. Keep your responses simple and make sure you address each issue the Examining Attorney raises. If he or she disagrees with you, you might receive a Final Refusal. This means you need to change the Examining Attorney's mind or you may have to appeal. If you get to this stage and you still don't have an attorney, please get one; there is a lot of strategy involved.

!

NOTE OF CAUTION!—A question of trust

The Examining Attorneys generally will not ask you about ownership or use dates. They trust you. Plus, you signed a declaration that says you can go to jail if you lie. But don't think you are getting away with something if you know there is a problem with ownership or use dates. Any dirty laundry you have will come out in any lawsuits you file later to protect the mark.

Publication About two months after your application is approved, it will be published in a book entitled the *Trademark Official Gazette*. You can't pick this book up at your local bookstore or over **AMAZON.COM**. You wouldn't want to. But there are a number of people who subscribe to it, like corporations who monitor their marks carefully, companies that have subscription services for doing this, and individual lawyers who are on the lookout for their clients. All these people, and anyone else, have 30 days from your publication date to file an extension to oppose, or to oppose.

Opposition Anyone can oppose registration of your mark for a number of reasons. They do not need to own a registration to do this. The most common reason is that they think your mark and product are too similar to theirs. But they can also oppose if they believe you are not the owner of the mark or if they think your mark is descriptive or generic. They can also oppose if they think you gave false or incorrect information to the Patent and Trademark Office.

If someone decides to oppose the registration of your trademark, the Patent and Trademark Office will notify you and provide a set of deadlines. If you have not started using an attorney at this stage, you definitely should start using one now. This is not something to fool around with. It is like a lawsuit.

Your opponent can request documents, which you are required to produce. They can summon you to be cross-examined and ask you all sorts of questions about your trademark, how you searched it, and your use. Many oppositions are settled in ways that allow both parties to continue using their marks. But you need to know what you are doing and be flexible in certain areas and inflexible in others.

Many oppositions are settled in ways that allow both parties to continue using their marks. But you need to know what you are doing and be flexible in certain areas and inflexible in others.

UPDATE YOUR MARKS

Brand owners sometimes update their logos over the years or even change their word marks somewhat. The Patent and Trademark Office knows this and will let you tweak your registration if it does not change the mark too much.

Hurray! Registration or Notice of Allowance

First of all, most marks are not opposed or threatened to be opposed. Assuming your mark was not opposed or that you have now weathered an opposition successfully and have not gone bankrupt in the process, one of two things will happen.

If you based your application on use or a foreign registration, you will get a registration. Go out and have a good time.

If you filed an "intent-to-use" application, hold off on the partying, you have more work to do. The Patent and Trademark Office will send you a Notice of Allowance. You have three years to file a paper saying that you are using your mark for the products and services covered by the Notice of Allowance. But to keep your application alive, you need to file an extension every six months. Once you have completed the rollout of your mark, then you can prove use.

If the Examining Attorney is happy with the papers you file, you will get your registration. And here's more good news—the Patent and Trademark Office (as well as the courts) will treat your original application filing date as the date your rights began throughout the United States. Anyone who started using a similar mark on the same or related goods after you filed your application will be out of luck if you go after them. This is known as "the constructive date of use."

PARTIAL PROTECTION

If you have trouble getting a federal registration, you can try to get a state registration in the states you do business in. This at least gives you some protection in those states.

Maintaining your registration

At this point, you have an asset that you can sell. You can use it to stop others from registering or using confusingly similar marks. And third parties will see your mark when they do searches. But you do need to continue to use your mark for the products and services listed in your registration or it can be attacked by someone. Don't warehouse or mothball your trademarks. Remember: use it or lose it!

The Patent and Trademark Office also requests that you file certain maintenance papers to keep your registration alive. Between the fifth and sixth years after your registration issues, you need to file a paper called a Section 8 Declaration of Use (because Section 8 of the Trademark Act talks about this). You also need to file more papers called a "renewal application" and another Section 8 Declaration at the 10th anniversary of your registration and every 10 years after that. You get an extra six months to meet these deadlines if you cough up some cash, but the Patent and Trademark Office will not remind you, so make sure the dates are on your calendar. In addition, after your registration is five years old, if you have used your mark continuously for five years for the covered goods and services, you can also file a special declaration. This declaration strengthens your registration even more. Lawyers love these and so do the courts. It is another layer of armor. This is known as a Section 15 Declaration (you got it—because Section 15 of the Trademark Act talks about this).

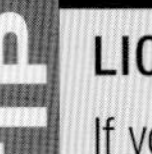

LICENSE AGREEMENTS

If you plan to go into the licensing business in a big way, it would be wise to have an employee dedicated to inspecting quality and acting as a go-between so that end consumers see a uniform image—whether the goods are made by you or by your licensee.

NOTE OF CAUTION!—Keep in the club

If you forget to file a Section 8 Declaration or your renewal, your trademark registration will be dead. You will still have your common law rights, but you're out of the registration club. You should try to get back in if you still use your mark.

STATE REGISTRATIONS

Each of the fifty states permits you to register your trademarks under their own laws. So does Puerto Rico. These are useful if you concentrate your business in one state or in a territory like Puerto Rico. But a federal registration is much stronger. It covers all these places.

AGREEMENTS THAT AFFECT YOUR TRADEMARK RIGHTS

Your trademarks won't exist in a vacuum. The reality is that they bump up against one another. Or others want to exploit them on merchandise. Or situations come up where you want to sell them or maybe you would like to mortgage them to get a loan. All of this is possible, but with any important asset, you need to be careful with what you do and don't do.

Coexistence agreements and consent agreements

During the search stage you might learn about another party's mark that is too close to yours. Or the Patent and Trademark Office might cite another party's application or registration to block your application. Or you might get a letter telling you to stop using your mark. You can

TIP

ASSIGNMENTS

If you are buying somebody's trademark but are not really buying the underlying business, don't get the false impression that you now own the brand. Sometimes people are willing to sell a trademark as part of a settlement where they have alleged your trademark infringes their trademark, or where you have approached them to work out a co-existence agreement. Unless you also buy inventory, design blueprints, customer lists, and other assets that are part of the "goodwill" which made the brand a trademark and not just a word, then all you are doing is getting rid of this person as a potential problem. That may not be a bad deal. But just keep this in mind if you do the deal that way.

consider approaching the other owner to work out a deal in these situations. The agreements that come out of these deals or settlements are generally known as coexistence agreements or consent agreements. They will help you get a registration if someone else has a similar mark. But they can be very dangerous if you do not know what you are doing. Also, expect people you approach to ask for money, a lot of it, and bargain from there. The Patent and Trademark Office will usually not accept a simple letter from the owner of a cited registration stating that he or she consents to the registration of your mark for particular goods and services. Instead, it wants to see an agreement with proof that both parties are taking steps to ensure no likelihood of confusion in the marketplace (see Appendix 2, pp.262–264 for a sample, but keep in mind that each case will vary).

Your trademarks won't exist in a vacuum. The reality is that they bump up against one another. Or others want to exploit them on merchandise. Or situations come up where you want to sell them or maybe mortgage them to get a loan.

License agreements

If you are lucky enough to own a brand that becomes popular, third parties might start approaching you to use the brand and all the imagery associated with it for goods and services they make. At one time, this sort of thing happened only in the entertainment and fashion fields. For example, the Walt Disney characters were heavily licensed. Now, there are licensing opportunities for all sorts of brands. Think of **MARTHA STEWART** as a brand. Even after all the issues that brand's founder had with the government, it is still a great brand that lots of companies would like to license.

This is where license agreements come into play. These are detailed agreements that grant third parties the limited right to use your brand (and that can include the word mark, logos, designs, and so on) for certain products or services in a certain geographic area for a certain period of time. In return, the other party will pay you some sort of royalty. You need to make sure you have the right to exercise "quality control" over the manufacture, design, and sale of the goods bearing your trademark. After all, it's your brand's image that's at stake when you let others use it.

NOTE OF CAUTION!—Quality control

This is serious business. If you fail to monitor the quality of the products, a third party can cancel your registration or attack your rights in your mark.

TIP

LICENSING A TRADEMARK

If you are thinking of licensing a trademark to use in your business—for example, to sell video games under a particular entertainment property or to sell fragrances of a well-known clothing designer—find out whether they have mortgaged their trademarks or have problems with debt. If they do, and if they go bankrupt, the court can actually terminate your license even if you prepaid a large royalty! You could be left holding the bag!

Assignments

If the time comes to sell your business, be sure to remember to think about your trademarks. If the buyer is buying all your assets and is going to carry on the business, he or she will expect you to sell your trademark to them. But a trademark is not just a word or other symbol. It has a reputation or goodwill associated with it. So any sale of your mark must also include the goodwill associated with the trademark. If the trademark has been registered, you should schedule the trademark on the sale agreement. Once that is completed, a simple assignment form can be recorded with the US Patent and Trademark Office. You should keep this in mind if you are buying a business as well—make sure you get the trademarks and associated goodwill.

!

NOTE OF CAUTION!—Intent-to-use applications

Do not assign intent-to-use applications except in the most limited of circumstances. Check this with your lawyer.

Other corporate changes also occur such as mergers and name changes

It is wise to record these with the US Patent and Trademark Office. Otherwise, there will be a great deal of confusion as to who really owns the mark when the time comes to file maintenance papers or if you want to sell your brand. This can be done electronically.

COPYRIGHT CLAIMS

If someone starts to use a design mark or logo that looks like yours, you might also have a copyright claim. This is so even if the person uses it for very different products or services.

Security agreements

In the same way that you can get a mortgage on your house, a bank will sometimes loan you money if you pledge your trademarks as collateral. If you get into this or any type of loan agreement involving your trademarks, make sure you consult with your accountant and attorney. Some banks use old forms that actually make you assign your rights to the bank. That doesn't make any sense! The bank should not own your trademarks because it does not know how to run your business. You wouldn't sell your house to the bank in return for a mortgage, would you? Then why would you sell your trademarks in return for a loan? Most of the banks have gotten their act together over the years.

ENFORCING YOUR TRADEMARK RIGHTS

Now that your brand is in the marketplace and is starting to take off, you want to make sure that no one else uses your trademarks. If your products or services are sold through a sales staff or agents, they will often be quick to let you know about possible problems. After all, they are counting on your brand to make money and they don't want other people getting in their way. When you go to trade shows, look to see what everyone is doing—not just your immediate competitors.

It's a no-brainer if someone literally copies what you are doing. These people are known as counterfeiters and the law provides some pretty clear steps for dealing with them. But what do you do when someone starts using

In the same way that you can get a mortgage on your house, a bank will sometimes loan you money if you pledge your trademarks as collateral.

a trademark that is identical to your own but for different goods or services? Or worse yet, someone starts to use the same or a similar trademark for competitive products and services? Sometimes these people are purposely trying to rip you off; at other times it is just a coincidence. Either way, they might be infringing your trademark rights and you can try to stop that. Sometimes you can even get money from them.

But remember, having a trademark right does not mean you own a word or design. Your trademark rights come from the use of the words, logos, design, packaging, and other symbols you have built up as your brand in connection with your product line or services. So unless your trademark is famous or an incredibly unique word, you are probably wasting your time getting upset about people using the same or similar marks in unrelated areas. Also, if you selected a trademark from the descriptive category, it is going to be a lot harder to stop other people from using similar trademarks. Even if they are competitors.

Cease-and-desist letters

The best proof that your mark is infringed is if consumers are confused. If customers start asking whether you have started selling the other product or if you have sold your brand to someone else, it means that you can start losing sales or the reputation of your business can be damaged. When that happens, it is definitely time to take action. The law doesn't require you to wait until you have suffered actual harm. It will also protect you if there is a

CANCELLATION

Once a mark has been registered for five years, you can no longer cancel it in the Patent and Trademark Office based on your prior rights. The courts will sometimes consider such a case.

likelihood that customers will get confused in the future. But what do you do? Here are some pointers before sending cease-and-desist letters:

- We stress that you should go to your lawyer to review the situation.
- Make sure that you have a protectable trademark. Is it registered? If so, have you been using it in the way it is registered and for the goods and services covered by your registration? Do you use your trademark in the region of the country where the other person does? If it is unregistered, have you kept good records of when you first began use, where you have been using, samples of use, and so on? If not, you might have trouble proving that you even own a trademark.
- Make sure that your trademark rights are earlier than those of the other party. The last thing you want to do is to send a cease-and-desist letter to someone who turns out to have been in the marketplace before you! Don't think you know the market. It's a big country out there and you would be amazed at how many people come up with similar ideas independent of one another. You can use investigators to learn more about the other party's use.
- Once you have all the facts, evaluate whether current or future customers are likely to believe that the other product comes from your business or that you are somehow associated with this product or endorse it. If so, the other person might be a trademark infringer. The enemy.

Check whether the other party has filed a trademark application. If so, get your lawyer to monitor the application. Once it is published, you can oppose it. If it is already registered, you can try to cancel it.

If your attorney agrees with you that you have a case, it is a good idea to start off with a cease-and-desist letter. Such a letter should clearly set out the rights that you have and what you have done with your trademark. Give the other side a copy of your registration if you have one, plus samples of your use or a selection of ads and such material. The goal here is to get the other party to stop as soon as possible. The more you show up front, the more likely it is that the other side will understand that you have a legitimate business and its activity is causing you problems. The letter should also explain why you think there is a likelihood of confusion. If you have genuinely confused consumers on your hands, say so (save the details for later). Give the other party a deadline for a response, say two weeks. If the case is really egregious, give less time and send the letter by overnight courier or certified mail.

If you do get a response, don't be surprised if it's nasty. It is very rare that someone just backs down. However, many responses to cease-and-desist letters end with some sort of olive branch. Like, "let's talk about it, even though we don't think there is a problem". That is usually a sign that you have someone willing to tweak a trademark or maybe even phase it out, given some time. If you are interested in settling, go for it!

If your brand is for luxury products or you have a popular brand, low-lifes will start to sell fakes.

If instead the infringer tells you to get lost or doesn't respond, you need to either sue or walk away. Don't walk away and then change your mind years later—the courts will not be very sympathetic to you unless the infringer makes changes to worsen the situation over that time period.

If you sue, expect mammoth bills from your lawyer (unless he or she is your cousin or is doing it on a contingency-fee basis). Also expect to have a lot of your time taken up with answering questions, giving copies of documents, being deposed, and basically being forced to spill your guts. Many people think this is a huge intrusion. But that is the way the system works in the United States.

NOTE OF CAUTION!—Stopping someone from using your mark

Remember, the Patent and Trademark Office can only stop someone from getting a registration. They cannot stop someone from using a trademark. So even if you spend a lot of money on an opposition and win, the person can still use the trademark. It requires a lawsuit to stop use. The court can also order the Patent and Trademark Office to abandon an application or cancel a registration. So if you know you have to stop use, it makes sense to sue. If you ask the Patent and Trademark Office to sit back and wait for an order from the court, it usually will.

If you sue, expect mammoth bills from your lawyer... Also expect to have a lot of your time taken up with answering questions, giving copies of documents, being deposed, and basically being forced to spill your guts. Many people think this is a huge intrusion. But that is the way the system works in the United States.

Watching out for other federal applications and registrations

Some of the large search firms have computer systems that go through the *Official Gazette* each week looking for similar marks. Some lawyers do this too. You can get reports with copies of new applications and recently published applications. There are also services that regularly scan through the Internet looking for ripoffs and also look at trade publications and other sources with unregistered trademarks. This can get expensive.

Whatever you do, make sure you look at these reports regularly, like once a week. Don't let them sit around or you can miss important deadlines. Besides, the sooner you put someone on notice about your rights, the more likely you are to settle.

Unfair competition

Even if your trademark is not registered, you can rely on your common law rights to stop others from using confusingly similar marks. Although your competitors can refer to your trademarks, whether registered or unregistered, to compare your products to their products, they have to be truthful. If your trademarks end up being used in a way that is disparaging or if there are false statements about them, you can stop that. This area of law is known as "unfair competition." Both state and federal laws have special provisions for this.

Stopping counterfeits

If your brand is for luxury products or you have a popular brand, low-lifes will start to sell fakes. These are known as counterfeit items. People who make counterfeit products can be subject to criminal prosecution in the United States. You also have these additional options available.

Customs recordals The US Department of Homeland Security Customs and Border Protection Office has a special branch that will let you record your federal registrations with them. Customs agents located at all the borders are on the lookout for counterfeit products coming in from other countries. Customs will seize and destroy counterfeit products and also assess fines against the importer. It is difficult to know how much product Customs really catches. But in recent years they have become much more aggressive. You want Customs on your side and they are there to help you.

Trade groups There are a number of coalitions and other trade associations that have banded together to go after counterfeiters in particular industries. This helps keep costs down. Also, counterfeiters do not pick on just one brand. They do a lot of different stuff. So this is one time when you want to get along with your competitors—you all have the same problem and we encourage you to get involved.

If you manage to get a domain name that corresponds to your trademark followed by ".com", you have a terrific edge in the marketplace.

TRADE NAMES AND DOMAIN NAMES

In addition to the various trademarks you use for your products and services, you will probably have other assets that can be protected. These include trade names and domain names.

Trade names

You can also develop rights similar to trademark rights in your trade name or business name. If you plan to have one main trademark for your products and services, you should consider selecting a trade name that is the same as your trademark. For example, Maidenform, Inc. owns the **MAIDENFORM** brand and Microsoft Corporation owns the **MICROSOFT** brand. These trade names will further help you identify and distinguish your business from the businesses of your competitors.

In certain trademark disputes, the additional umbrella of protection provided by trade names turns out to be very useful. They can give you rights that may predate your first use of your trademark. They can also help you establish your reputation, since your business name is likely to show up in places where your brand may not—for example, in articles and press releases. By bundling together your trade names and trademarks, you can strengthen your overall portfolio. If you incorporate under a name in a certain state, or make a fictitious name filing, that alone does not give you protectable trademark or trademark-type rights. Just like trademarks, your trade names have to be used in the marketplace.

By bundling together your trade names and trademarks, you can strengthen your overall portfolio.

Internet domain names

Domain names have quickly become an extremely useful tool to help promote brands. Websites at the addresses for these domain names can be used to sell product or give potential consumers information. The Internet is like a big billboard showing up on everyone's desk, laptop, and PDA. Just think of **AMAZON.COM** or **BARNESANDNOBLE.COM** in the field of bookselling. Without even making use of a search engine, we intuitively visit these sites by typing the brand name followed by ".com". If you manage to get a domain name that corresponds to your trademark followed by ".com", you have a terrific edge in the marketplace.

Therefore, when you are selecting and searching your trademarks, you should simultaneously be finding out whether the corresponding .com domain name is available. If it is not, perhaps you can add a generic word to it, which will still be of great value. For example, Delta Airlines has **DELTA.COM**. But what about Delta Dental? You can find them at **DELTADENTAL.COM**. No problem.

It is very easy to check the availability of domain names yourself. For example, visit www.networksolutions.com. If the domain name you would prefer or even one you think would work well is not available, it might be for sale. First, check whether it is active—there is a website at www.centralops.net/co/ that will help you figure that out. If it is not active or if when you enter the address the page indicates it is for sale, you can contact the owner. Information on the owner and his or her contact particulars are also at www.networksolutions.com.

In the early years of the Internet, stories about domain name owners selling domain names for huge sums of money were widespread. Every once in a while this still happens. But in general people have gotten a lot more realistic. You might come up against someone who is sensible and reasonable—it is definitely worth a try.

You have also probably heard a lot about domain name pirates. These are people who register the names of famous personalities or businesses or variations of these names. When you are adopting your brand, you should watch out for pirates. You should register the domain names you want to use as well as similar ones that your customers might type in intuitively before you make any announcements about your new brand. But it is impossible, and certainly not financially sensible, to register all conceivable domain names.

When adopting your brand, watch out for pirates. You should register the domain names you want to use as well as similar ones that your customers might type in intuitively before you make any announcements about your new brand.

If you end up having a popular brand, expect to encounter pirates who get domain names similar to your marks. How do you deal with them? Some companies go after everyone, whether the website is active or not. Other companies selectively and methodically pursue the pirates who have active websites or who are trying to sell the domain names. We think this latter approach makes a lot more sense. You certainly do not want to encourage the pirates. In fact, there are now US laws and international agreements that can force a pirate to transfer the domain name registration to you. For information on the international system, you can visit www.arbiter.wipo.int/domains/index.html. It would be wise to use a trademark lawyer to start one of these proceedings, since there are certain do's and don'ts.

CHAPTER 2 rights of publicity

WHAT IS A RIGHT OF PUBLICITY?

Turn on the television and there is **CATHERINE ZETA-JONES**. But you're not watching **"CHICAGO."** Instead, there she is promoting **T-MOBILE** telephone services.

From the theater to the ballpark and from old VCRs to the Internet, we have access to endless drama, sports, motion pictures, television programs, and music. Along with such entertainment comes the immediate recognition of celebrity names and faces. Even if we have never met them or seen them in person, we feel like we know celebrities. For over a hundred years, our country has had a love affair with stage actors and actresses. And with the advent of television, an orgy began. We became obsessed with what they did in their spare time, who they married, and yes, what types of clothes they were wearing or products they bought. It was only a matter of time before celebrities began to use their status to hawk everything from antidepressants to popcorn and to become spokespersons for a wide range of brands. And it is not just established brands. Celebrities use the weight of their fame to launch their own brands. Sometimes the brands are unrelated to the celebrity name—for example, Elizabeth Taylor's **WHITE DIAMONDS** brand perfume. More often, a celebrity will create a brand to exploit his or her

And look, there's **TIGER WOODS**. But he's not out on the green. Instead, he's promoting **NIKE** shoes. And wait a minute, isn't that **BEYONCÉ** drinking a **PEPSI**? Why isn't she singing?

recognized name, whether it be **NEWMAN'S OWN** for Paul Newman's line of food, **KENNY ROGERS ROASTERS** for the singer's chicken stores, or **MARTHA STEWART** for the domestic diva's line of housewares.

Whatever brand a celebrity chooses to adopt or endorse, the brand can eventually eclipse whatever it was the celebrity originally did to earn celebrity status. Admit it, when you hear **GEORGE FOREMAN**, don't you think of the brand for the grill first?

But what happens when your name or your photograph or your voice is used to promote someone else's product or service when you have nothing to do with that product or service? And what happens if someone dresses up to look like you and promotes a product in an ad? And what about fake voices mimicking your voice so that it sounds as if you are endorsing a product on a radio commercial? Most of us don't need to worry about this

"Rights of publicity" are designed to protect unauthorized commercial uses of your name, your "likeness," such as a picture or drawing of you, and your voice.

because our name or picture or voice has no "drawing power." But if you have "drawing power" this can become an issue. Being a celebrity is a big deal. Being a celebrity who gets paid money to endorse other people's products or to start up a brand is big business. And if you have the good fortune of being a celebrity, or if your name, or face, or voice has drawing power, you need to understand the rights of publicity, how to protect your rights, and what to do if they are misappropriated. And, if you have a product or brand that you would like to have endorsed by a celebrity or someone with drawing power, you need to make sure you do not violate his or her rights of publicity.

HOW DO RIGHT OF PUBLICITY LAWS WORK?

The trademark laws protect celebrity names like **MICHAEL DOUGLAS** and performing names like **J.LO** when used as a trademark for specific goods and services like acting or on CDs. Even before celebrities started using their names as trademarks, people used their personal names in their businesses as a trade name or trademark so consumers would know who they were dealing with, like **J. P. MORGAN** or **R. H. MACY**. However, in commercial or business settings, people often use their names, but not as a technical trademark or trade name.

For example, even though we, Fred Mostert and Larry Apolzon, are using our names as authors of this book, we are not using our names as trade names or trademarks. These are just our personal names as authors.

While the trademark laws have some provisions to stop others from using personal names to falsely suggest an endorsement of, or association with, their goods or services, there is no formal registration process for your name per se. If we were to set up a business for publishing books, like Fred Mostert PLC or Larry Apolzon LLC, and hand out business cards and use our names on stationery, our names would also be trade names. At that point, the trademark laws would start to provide some additional protection to us. Then, if Fred were to start up a book imprint to publish books like **FRED MOSTERT & COMPANY**, or if Larry were to adopt a nickname like **LARRY'S PUBLISHING** and use these as trademarks on books, we would have trademarks that we could register in the Patent and Trademark Office. At that point, a full array of trademark rights would kick in.

While the right of publicity has existed for some time, it is a relative newcomer to the world of intellectual property.

While the trademark laws and laws of unfair competition protect your name or nickname when used as a trade name or trademark, these laws are not nearly as helpful in protecting the use of your personal name, or your picture, or your voice. So what do you do if your name is not also a trademark or trade name and someone starts to use it, or a picture of you, or your voice in ads to peddle their products or services?

Under many state laws, the rights of publicity can last for as long as 100 years after the death of a celebrity.

This is where "rights of publicity" come into play. They are designed to protect unauthorized commercial uses of your name, your "likeness," such as a picture or drawing of you, and your voice.

While the right of publicity has existed for some time, it is a relative newcomer to the world of intellectual property. Except for some unfair competition provisions in the trademark laws, the protection of rights in your name or likeness or voice is largely left up to the individual states we live in and where people do business. The majority of states have some type of protection. A few, including Iowa, Oregon, South Dakota, and Wyoming, offer no protection. In the states of California, Illinois, Indiana, Kentucky, Massachusetts, Nebraska, Nevada, New York, Ohio, Oklahoma, Rhode Island, Tennessee, Texas, Utah, Virginia, Washington, and Wisconsin, the state legislatures have passed laws to protect these rights. The remaining states rely on prior court decisions to guide them. And in a few other states, like California, Florida, Indiana, Illinois, Nevada, Ohio, Tennessee, Texas, and Wisconsin, there are both written laws and legal precedents that will protect you.

A few of the states, such as Indiana and Oklahoma, have very favorable laws, regardless of where you live. You do not even need to live in Indiana or ever even set foot in Gary, Indiana, to protect your right of publicity. And California has good laws if you live there. Under many state laws, the rights of publicity can last for as long as 100 years after the death of a celebrity.

TIP

BE CAREFUL WITH "DECEASED PERSONALITIES"

In California, heirs and other claimants of "deceased personalities" can register their rights and use the laws of that state to go after people for certain uses of the name long after the celebrity is dead. Texas, Nevada, Indiana, and Oklahoma also have registries. This can come as a real and unpleasant surprise to companies who might use the name of a dead author or personality in an advertisement or in some other commercial venture. Unless a personality has been dead for over 100 years, you need to do some research to find out whether their heirs could have a claim against you if you use their name or likeness in a commercial venture. Plus, you need to find out whether the name has been used as a trademark. If it has, you need to make sure you do not violate any trademark rights.

The right of publicity laws grew out of the "right of privacy" laws. These laws are known as "the right to be left alone," which stops others from using your name or pictures of you for someone else's advantage, intrusion, disclosure of private facts, or portraying you in a false way.

DEBUNKING THE MYTH—Public recognition

The "right of privacy" does not help celebrities and other personalities like politicians very much. Celebrities are pictured all the time in newspapers and on television; and you hear quips about them on the radio on all sorts of personal issues beyond their most recent movie, television episode, or record release. The courts have come to recognize that this type of use, if true, is fair play when you are a celebrity. Also, if you are not a celebrity, but get involved in some sort of newsworthy event, like a car accident, the media is allowed to use your name and picture and talk about the event, as long as what they say about you is truthful and relates to the event. But whether you are a celebrity or are momentarily in the spotlight, or are even just minding your own business, that doesn't give anyone a license to use your name or picture to promote car insurance or a tow service. If that happens, and if the state this happens in protects your "right of publicity," you can sue the insurance company or tow service and stop the use, and maybe even get them to pay you some money. In fact, in some states, like New York, it is a misdemeanor to do this.

OTHER LAWS PROTECTING NAMES

Trademarks

Relying on the rights of publicity laws set up by the various states to stop commercial uses of your name or likeness can be very useful. But judgments you get might be limited to the state you get a judgment in—it is not nationwide. While a judgment in one state can be enough to shut down a national campaign because of the pervasive nature of media, this is still a hassle if you have a nationwide problem. As a result, people often turn to the federal trademark laws when they believe their "right of publicity" is violated. In these cases, they allege that the commercial use of their name or likeness is a false endorsement or unfair competition. But unless your name was used as a trademark or trade name before this, the trademark laws will not generally help. Here's an exception though. If you are a celebrity or if your name has some type of drawing power, the federal trademark laws can go to work. For example, Bette Midler was allowed to sue a major car company for a television commercial in which a "soundalike" voice sang "Do You Want To Dance?"

If you would like your name or likeness to be protected against unauthorized uses, or if you want to license the use of your name and also allow your heirs to control the use of your name after you die, then you should turn to the trademark laws to protect your rights. You can always fall back on rights of publicity, but it is much better to also have trademark rights in your name. This will give you some very powerful rights.

Relying on the rights of publicity laws set up by the various states to stop commercial uses of your name or likeness can be very useful. But judgments you get might be limited to the state you get a judgment in—it is not nationwide. While a judgment in one state can be enough to shut down a national campaign because of the pervasive nature of media, this is still a hassle if you have a nationwide problem.

...your name does not even need to be famous and you do not need to be a celebrity to take advantage of the trademark laws.

A trademark in your name or a logo with your portrait will last for as long as you or a licensee uses it. Think of **ELVIS PRESLEY**. That name as a trademark will probably survive all of us and our grandchildren's children too!

If you plan to use your name or license the use of your name to others for specific products and services, you can file an intent-to-use trademark application. And if your name is already in use, you can file a use-based application. This is the best way to get nationwide protection in your name. In fact, your name does not even need to be famous and you do not need to be a celebrity to take advantage of the trademark laws if you actually use or license out the use of your name as a trademark.

If you do file an application though, remember that eventually you need to prove that you use your name as a trademark. You would think that the prominent display of an actor's name on several different movie posters, where he or she stars in the movie, could get a trademark registration for "acting services." But some Examining Attorneys in the Patent and Trademark Office disagree. On the other hand, if you have a website where your name is prominently displayed and your movies are listed, Bingo! You will probably be granted a trademark registration for your name. The same is true if you have a flyer or fact sheet displaying your name.

Here is another advantage of using the trademark laws for famous names. You can rely on the federal antidilution laws to stop unauthorized

TIP

THINK BEYOND PROTECTING JUST YOUR NAME

You can also register a sketch or portrait of your face or a silhouette as a trademark. For example, a company of the rap artist and entrepreneur **SNOOP DOGG** is registering his profile in the clothing and music fields. There is no doubt that this type of protection will give the broadest protection, as a trademark by definition is a commercial designation and the courts, both federal and state, will protect that. You can go one step further and even register the copyright in any design mark that uses your likeness.

trademark uses of your name that have nothing to do with the services you cover in your registration. Remember from the chapter on trademarks that infringement occurs if your mark or a similar mark is used in connection with the same or related goods or services that you provide. But the infringement laws do not help if the goods or services are not related. This is where the dilution laws can help. If your name is famous and you also have trademark rights, you can allege that a third party's use of it dilutes your trademark rights and you can stop the use.

!

NOTE OF CAUTION!—Dodging the bullet

These laws have exceptions, so don't think this is a magic bullet. But it can give you a lot of leverage against people who are unfairly exploiting your name. In fact, in recent years we have had a parade of celebrities showing an interest in registering their names as trademarks to help stop unfair uses of their names.

Your name is one of your most important assets, especially if you have worked your whole life to build up a reputation. But you would be amazed at how many celebrities, personalities, and designers have given third parties rights to own or register their names as trademarks. Don't confuse giving up your name with licensing it out. The famous late designer Halston gave away his **HALSTON** name as a trademark. You shouldn't let that happen to you. Once you give up trademark rights in your name, your ability to license it out to others can be limited or even prohibited. Further, you no longer have

Once you give up trademark rights in your name, your ability to license it out to others can be limited or even prohibited.

an asset to sell when you retire. The problem is that many young designers and celebrities love to see their names on labels and in lights—what an honor! And some employers and other companies who see you as a "hot property" will take advantage of this and leverage that opportunity in order to get 100 percent ownership rights to your name. If this happens to you, do the best you can to retain ownership of your name as a trademark. Offer to give your employer an exclusive license. You can even agree not to license out your name to someone else for a lengthy period of time or limit types of licenses you will give.

In some cases, you might be able to divide the trademark right in your name so that you have it for one business and another party has it for another. However, this weakens your rights. But if that is not good enough and you really want to do a deal, you should review this carefully with a lawyer and come up with an agreement that gives you the right to get your name back after a certain period of time, if you pay a certain amount. You should also include an option clause that gives you the right to buy back your name for a certain amount before they are allowed to sell the trademark rights in your name to someone else.

Make sure you get as good a financial deal as you can. Giving away your name is a once in a lifetime opportunity. We hate to see it happen, but if you are forced to do it, make it the best deal of your life.

Don't confuse giving up your name with licensing it out.

Domain names

If you are headed for stardom or some sort of special public recognition, you should immediately register your name and any domain names you might want to use or don't want others using. If you have a common name, it might already be registered by someone else. You might be able to buy it from them before you become well-known. Don't get too bogged down in gobbling up every conceivable variation of your name. Pirates will always figure out new ones to use. Instead, focus on the domain names you really need like firstnamelastname.com or yourperformingname.com or .biz People in television sometimes get a domain name from the country of Tuvalu that ends in ".tv," which is pretty cool. See the trademarks chapter on what to do if a pirate registers your name as a domain name in bad faith.

Copyrights

If there is a recording of your voice that really defines you, or a photograph that everyone loves to see, you should enlist the cooperation of the owner of the copyrights in these creations to keep them out of the hands of people who want to use them commercially. Remember, just because it is your voice or your photograph does not mean that you own the copyright in it.

If the copyright owner does not want to assign the copyrights to you, then you should at least get him or her to agree to cooperate if you ever need to rely on the copyrights in that creation.

The recording studio and photographer will own all, or will at least be a co-author with you. It would be best to get an assignment of these copyrights. Then, if someone else uses that recording or picture commercially, you will have copyrights on your side. If the copyright owner does not want to assign the copyrights to you, then you should at least get him or her to agree to cooperate if you ever need to rely on the copyrights in that creation. And whatever you do, make sure that any photographers or recording studios who capture your voice or likeness agree not to use it commercially except in specific ways that you authorize in writing beforehand.

Remember, just because it is your voice or your photograph does not mean that you own the copyright in it. The recording studio and photographer will own all, or will at least be a co-author with you. It would be best to get an assignment of these copyrights.

WHAT HAPPENS IF I THINK MY RIGHT OF PUBLICITY HAS BEEN VIOLATED?

As we say in connection with any violation of your intellectual property rights, you should immediately consult an attorney who is knowledgeable in this area of the law. Many trademark attorneys have become sophisticated in this area since so many celebrities have registered their names and likenesses as trademarks over the years, and because trademark claims are also linked to right of publicity claims. Initially you might think that a celebrity would be flattered to have his or her name associated with someone else's product or service. But what if that service is a bad one? Or what if that celebrity has already entered into an agreement to endorse a competitive product or service?

It is extremely important to make sure you get appropriate permissions before you start to use a name, a picture, or anything else associated with a person to promote your products or services.

It is not uncommon in these types of cases for the defendant to allege that the Constitution gives the right to use the name or likeness. After all, the First Amendment is generally considered a little more important than trademarks and other commercial rights! The courts have caught on to this and are pretty good at deciding whether or not a use is protected under the First Amendment or whether it is a violation of the rights of publicity.

Another type of defense that can come up if you are relying on the trademark laws is that the use of your name is a "fair use." In trademark law, as with the parody defense in copyright law, the fair use defense gets thrown around more than a baseball in the World Series. In such cases, the courts are asked to decide whether the name has been used in a way that describes the goods or services and has been used fairly. For example, the office furniture brand **HERMAN MILLER** was allowed to use the name and likeness of the furniture designers Charles and Ray Eames to promote reproductions of furniture that was originally designed by them. In other types of "fair use" cases the courts decide whether the personal name really needs to be used to make any sense of the product or service.

If you decide to pursue a right of publicity or a trademark claim in your name or likeness, be prepared for a roller-coaster ride that can be very expensive. In recent years, celebrities and estates of celebrities have become entangled in disputes involving rights of publicity where decisions were first made in their favor and then reversed by upper courts. In some instances, defendants were even awarded large sums of money and other cases grew out of them. Communicate with your lawyer about what you really want out of the case. Do you want money? Will you be happy if the use is stopped? It's a lot easier to stop someone than to get them to pay you a lot of money. Once money is involved, you can get yourself into a pretty expensive poker game.

Communicate with your lawyer about what you really want out of the case. Do you want money? Will you be happy if the use is stopped? It's a lot easier to stop someone than to get them to pay you a lot of money. Once money is involved, you can get yourself into a pretty expensive poker game.

CHAPTER 3 copyright

WHAT IS COPYRIGHT?

Stop reading for a moment. Look up. Look around. What do you see? Are there **POSTERS** or **PAINTINGS** on your walls? Are you wearing a piece of **JEWELRY** with some **ORNATE INSCRIPTIONS**? Do you have a fancy **TATTOO**? What about the

The truth is, we are surrounded by paintings and other visual art, writings, software, music, and other creations where someone has expressed themselves in some sort of medium. And all of these expressions may be protected by copyright from the moment they were created, whether or not they are registered in the US Copyright Office.

Large media and entertainment companies, advertising agencies, magazines, and publishers rely heavily on copyright protection. If you are a writer, a visual artist, a website builder, a graphic designer, a songwriter, a photographer, a musician, a movie producer, a comedian, a performer,

PATTERN on your clothes—is there anything special about it? And if you are watching **TELEVISION** (we discourage multitasking while reading our **BOOK**, by the way), what about the television program? Is it a sitcom? A **MOVIE**? The nightly news?

a choreographer, or any other creative person, you need to know that your creative works can be automatically protected under the copyright laws the moment you express them on paper or in some other fixed format. But you also need to make sure other people know it's your creation and be aware of what you can do if someone copies it or uses it without your permission.

If your "creation" ends up not being protectable under the copyright laws, you might be able to protect it with a design patent or a utility patent. Or if you use and promote it in a certain way for a long time, you might even be able to protect it under the trademark laws.

DEBUNKING THE MYTH—What isn't covered

Even though the list of what copyrights protect seems to cover everything under the sun, it does not. Copyright protection extends to original works of authorship but not to ideas, "useful articles", simple geometrical designs, typestyles, short words or phrases, or titles. It is not always easy to tell whether an article is just useful or has some sort of artistic merit to it, allowing the copyright laws to kick in. Most clothing designs and even many jewelry designs are not protected under copyright laws. At the same time, elaborate fabric patterns used on dresses, or even couches, and highly ornamented jewelry pieces can be protected under copyright.

Express yourself!

Take your ideas and create something real with them! We cannot stress enough how important it is to get your ideas out of your head and onto paper or your hard drive. The moment you use your own ingenuity to create an original story, a poem, an article, a painting, a drawing, a song, or even software, you automatically have copyright in your creation. And as usually happens, once the idea is out of your head and onto paper or into your laptop, other ideas start to flow and the work begins to grow and develop. Your new creation might not be a Picasso painting or a David Mamet play or a John Lennon song, but it's your copyright to develop, keep private, sell, or exploit however you wish. With some talent, ingenuity, marketing, and a lot of luck, you might have a valuable asset, and if someone copies it or uses portions of it, they will have violated your copyrights and they can be in a lot of trouble.

We all have ideas. It is getting them down on paper that's tough.

Because you need to get your ideas down in some medium, it's not a good idea to tell your friends or family or anyone else about what you have in your head. It's tempting to tell people about your ideas because new ideas can be very exciting. But it is better to keep them to yourself until you get them in a fixed form.

NOTE OF CAUTION!—Sharing your idea isn't enough

Telling someone about something does not give you any copyrights (unless a tape recorder is running all the time you are speaking and you keep the tape). If someone takes one of your ideas and writes it down before you do, you are out of luck. We all have ideas. It is getting them down on paper that's tough. And it's that sweat equity and creative spirit that the copyright laws are there to protect.

Your creations need to be original

In addition to getting your ideas onto paper, you need to make sure that your creations are original. Just as no one can copy from you, you cannot copy from someone else if their work is still protected. Make sure your creations represent your own fresh and unique interpretation of the subject. Obviously, we are all influenced by the world around us. So anything you create will inevitably have influences from the outside world, but there is a big difference between that and illegal copying. Even though your creations need to be original, there might be situations where two people who don't know each other or have anything to do with one another might independently come up with creations which are almost the same. It's a big world out there and this sort of stuff happens. That's fine. In such cases, each person will have copyrights in their individual work.

What else do the copyright laws protect?

Not only do your copyrights give you the right to stop others from reproducing your protected work—you can also stop others from creating derivative works or spinoffs from your works. They need your permission, just as you need permission to create derivative works. You also have the right to stop others from selling, displaying, or performing your works without your permission.

So what does this mean? For any type of creation you have, selling an original painting or piece of jewelry or sending out a copy of your book manuscript or TV show proposal does not mean that you give away your copyrights. The other person might own the physical object with your work, but you still own the copyrights in the actual creation. Copyrights are set up so that you can pick and choose which rights you want to give up or license. After all, they are known as "copyrights."

DEFINITION—Moral rights

If you are a visual artist, you have the rights to have your name associated with your works. These are known as "moral rights." You can stop others from associating their name with your artwork. In addition, you can stop others from distorting or mutilating or even modifying your creations if it would be prejudicial to your reputation. In some cases, you can even stop the destruction of your artwork, even if you sold or gave away the physical object. You should retain a copy, digital or otherwise, for record purposes of your work if you sell the physical object it is embodied on.

How long do my copyrights last?

The answer to this question depends on when you created your work and who owns it. The copyrights in any work created after January 1, 1978 that you create on your own are protected for the rest of your life, plus seventy years. If you have created the work with someone else, the copyrights last for seventy years after the life of the last surviving author.

If, however, the work you create is as an employee and is part of your job, this is a work for hire. Your employer owns it. The same is true if you create a work as part of a magazine, a motion picture, a translation, a component of a larger work, or certain other types of specific ordered works and sign an agreement saying the work is for hire. A work made for hire lasts for 95 years from the date it is first distributed, sold, or offered for sale, or 120 years from the year of its creation, whichever occurs first.

The best way to have proof that a work is yours and exists as of a particular date is to register it with the US Copyright Office.

PROTECTING YOUR COPYRIGHT

Even though your works and creations have automatic copyright protection from the moment you get them out of your head onto or into a tangible medium, you should develop good habits to keep track of what you create and when. It is amazing how many copyright cases turn on formalities and on proving that you are in fact the owner of the work in question. Another common issue is proving exactly what you created at a particular moment in time. Proving that you are the rightful owner of a particular work often forms the nub of a copyright case. Surprisingly, not as many copyright cases as you might think hinge on whether there is infringement.

You might wonder how this can happen when we just told you that copyright automatically exists with the owner upon creation. If you created it, it's yours, right? Yes. But from long experience we can tell you that from the perspective of ending up in a lawsuit over your copyrighted works, it is crucial that you are in the best possible position to prove that you are the owner of a particular work and what exactly that work was at a particular time. If you've created it, it's yours. But go the extra mile and make sure you can prove it's yours and when you created it.

The best way to have proof that a work is yours and exists as of a particular date is to register it with the US Copyright Office, preferably as soon as it has been created. But here are some other steps to take as part of the process.

▶ Keep good records of your creations

To make sure you have proof of when you create your works, you should get into the habit of personally signing and dating the bottom of each page or document you create when you create it. If you are working in an electronic medium, print out copies and sign and date each page. Keep these in a safe place. When you make revisions, put those dates down as well, and sign again. Keep these versions in a safe place too.

If you are a really creative type and turn out a lot of work, you can make up a stamp as follows:

> *I, Mary Doe, confirm that*
> *I am the author of this*
> *design, which is an original work*
> *Signed* ______________
> *Date* ______________

If you are extremely concerned about someone copying your work, you can have someone else verify your signature. To do it properly, the verification should take place on the same date that you complete the creation of the work. If you do not have access to a witness, you can get your signature notarized instead.

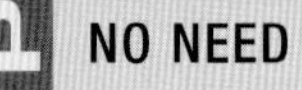

TIP

NO NEED TO WAIT

You don't need to wait until your work is finished and ready to be published to get your copyright registration. You can get registrations for various versions of your work at different stages of its development.

▸ Use a copyright notice

When you go to sell your work or start sending out copies to publishers or before you show it to anyone, even your mother or spouse or significant other, you should definitely place a copyright notice on all your copies. It lets people know you own the copyright. You mean business. You do not need a copyright registration to do this. The copyright notice should show the word **COPYRIGHT** or ©, followed by your name and the year you published it. If an earlier version was published, you can include all the years that you published the various versions to show the history. You do not need to include the year on certain items, such as greeting cards, jewelry, and toys.

Display the notice where people can see it. Don't be shy. It's also a good idea to add "All rights reserved" or "All unauthorized use is prohibited." It's like one of those "Beware of the Dog" signs. Here's an example:

© Mary Doe 1999, 2007. All rights reserved.

▸ Get a US copyright registration

The best way to have proof that as of a particular date your work existed in a particular form is to file a copyright application with the US Copyright Office. The United States is one of the few countries where copyrights can be registered. Even people in countries outside the United States get these

The United States is one of the few countries where copyrights can be registered. Even people in other countries get these to help prove their rights in their own country.

to help prove their rights in their own country. If you are a US resident, you also need a copyright registration if you want to sue someone for infringement. And having the registration in hand before someone infringes one of your rights can also help you get automatic money damages (see p.147). Visit www.fromedisontoipod.com to look at an example.

DEBUNKING THE MYTH—Don't bother with this old hat trick

There is an urban legend about proof of authorship that has been going on for years: placing your work in an envelope and mailing it to yourself. You have been told that the official date on the postage stamp will then be proof of when your creation was conceived. What a strange practice! What are you supposed to do with this thing? Wait until you are in front of a judge and jury to open it for the first time after you get it back? How do they know that you didn't tamper with the envelope? And if you open the envelope before you are in front of a judge and jury, in front of a friend, or your lawyer, then they are only a witness to the fact that you opened the envelope. You could easily have tampered with it before you showed it to them. So forget about this option—it is a waste of postage.

Forms The Copyright Office has a great website at www.copyright.gov. There are different types of forms to use, depending on the nature of your work. Each form includes instructions. "TX" forms are for written works such as books, computer programs, and so on. "VA" forms are for paintings, sculptures, architectural drawings, fabric designs, and other types of visual

Just remember there is no free lunch in the Copyright Office. You will need to demonstrate that there is some artistic expression involved before they will register your work.

works. "SR" forms are to be used for sound recordings. "PA" forms are to be used for songs, plays, motion pictures, and other audiovisual works. "SE" forms are for serialized works like newsletters, magazines, and newspapers. Whichever form you use, you need to be extremely careful to provide the correct information. The Copyright Office website is also a source of numerous items they call "circulars," which provide very useful information on many aspects of copyright law.

If you want to make sure that your copyright registration is a real asset that you can sell or assert in a lawsuit, you should visit a lawyer who litigates copyright matters or files copyright applications before you file your first application. He or she can walk you through the procedures and teach you the basic rules. Then you can go off on your own and start registering your copyrights. If you have an established business that needs lots of copyright registrations, or your copyrights are the most important asset you own, you might want to have one of your employees learn how to do this or get a special deal in place with your lawyer.

Samples The Copyright Office also requires that you submit copies of your work so that everyone knows exactly what you are registering. You need to submit one or two copies, depending on whether the work has been published and also whether it was published in the United States or outside the United States. Make sure to keep a copy of the sample and the completed form that you submit.

Examination After you file your copyright application, it is examined in the Copyright Office. This can often take six months. If the Examiner does not have any questions, your copyright registration will issue in the form of a copy of your original application form with a registration number, which is sent back to you. Its effective date will be the date the application was received.

If, however, the Examiner believes that the work does not exhibit enough originality or artistic expression, or if there are inconsistencies in your application, he or she will write to you explaining why the registration is not being made. You need to respond to the issues satisfactorily within 120 days or the application will be abandoned.

Some issues raised by the Copyright Office can be readily fixed. These include where you accidentally leave off a piece of information such as the date of first publication. Other inquiries can raise very serious issues. These include situations where the Examiner says that your work is only a commonplace functional article. Or he or she might say that it lacks sufficient "artistic merit." This comes up in different situations but includes refusals to register certain fabric designs or jewelry items because they have only geometrical designs.

Your new work may also be derived from an old work that was never protected under copyright or where the copyrights have expired.

Examples of works not subject to copyright
Examples include common or short phrases, titles and slogans (the trademark laws might be able to help you out in these situations). Also, familiar symbols or designs, lettering, or coloring cannot be registered in the Copyright Office—for example, a fleur-de-lis design. A mere listing of ingredients or contents and even blank forms, blank checks, or works that consist entirely of information that is common property containing no original authorship can be rejected. However, each case will vary. If you have a form with artwork on it, you can register the artwork if sufficient artistic merit is shown. Just remember there is no free lunch in the Copyright Office. You need to demonstrate that there is some artistic expression involved before they will register your work.

Issuance If you get all the questions answered, your copyright registration will issue after the Copyright Office makes the requisite changes in your application. As noted above, the registration will be a copy of your application form with the registration number and issue date filled in on the upper right-hand corner. Congratulations! We're sure you will have plenty more creations where you will want to register your copyrights.

STAYING OUT OF TROUBLE

If someone copies a portion of your work, you may want to go after them for copyright infringement. But this is a two-way street. If your work is inspired by someone else's work, that's not copying—or is it?

CORRECTION

Page 135

Incorrectly states that the paintings of Grandma Moses are in the public domain. The sentence should read "For example, the paintings of Leonardo da Vinci are in the public domain." We apologize and hope this causes no inconvenience.

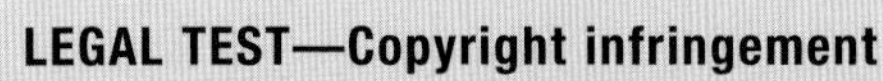

LEGAL TEST—Copyright infringement

The Copyright Office's examination only evaluates the sample of your work; it does not compare your work with any earlier work. This is where issues about whether or not your work is originally created as opposed to being another work in the "public domain", or whether the copying is a "fair use" of the work get involved. If you are only inspired by a work and get ideas that you put down on paper, that is fine. But if your work takes portions of another work, or is somehow based on its expressive content as opposed to unprotectable components such as the ideas or facts it contains, then you need written permission to use it unless that work is in the public domain or your use is a fair use.

Public domain

Works for which protection has expired, or never existed in the first place, are in the public domain and are said to comprise the body of all creative works and other knowledge considered part of our cultural heritage. For example, the paintings by *Grandma Moses* are in the public domain. So are the poems of *Emily Dickinson.* Therefore you can copy these works, create sequels or spinoffs, or do whatever you please with them (you should still give credit where due). But the more recently created paintings of *Andy Warhol* and *Picasso* or the songs and recorded performances of *Elvis Presley* are not for your use. You need written permission if you want to base any of your works on these works.

Your new work may also be derived from an old work that was never protected under copyright or where the copyrights have expired. These old or unprotected works are in the public domain. But be careful, since copyrights last for a long time and even works you think are in the public domain may still have certain aspects protected. The book "**THE WIZARD OF OZ**" by *L. Frank Baum* is in the public domain. But the later classic *Judy Garland* film by *Warner Brothers* contains significant new material that is different from the book. The additional material in the *Warner Brothers* version, and the movie itself, are still protected by copyright laws. So if you are basing your creation on someone else's work, make sure you have their written permission or be sure it is in the public domain.

NOTE OF CAUTION!—Outside the United States

Remember that laws vary from country to country. A work in the public domain in the United States may not be in the public domain overseas or vice versa. If you plan to distribute your work outside the United States, unless you are absolutely certain that the work you copied is in the public domain around the world, you need to check on a country by country basis. Even big entertainment companies have tripped up on this one.

Fair use

The copyright laws are filled with all sorts of exceptions as to why someone should be allowed to copy your work. But it boils down to this. Copies of your work or portions of your work can be used if it is for criticism, comment, news, reporting, teaching, scholarship, or research where the use will not be a substitute for your work or interfere with the potential market for, or use of your work. There have been many expensive lawsuits fought over this issue. People throw around the term "fair use" to justify what they are doing more than the American flag is flown on July 4. Consult your lawyer if you are doing this.

The basic factors that you need to consider include:

- The purpose and character of the use, including whether it is for commercial purposes or nonprofit educational purposes.
- The nature of the copyrighted work you are copying from.
- The amount and substantiality of the portion used in relation to the copyrighted work as a whole.
- The effect of the use upon the potential market for or value of the copyrighted work.

DEBUNKING THE MYTH—Fair Use

Somewhere along the line, certain people got the idea that a given number of words or lines from a literary work or bars of music from a song can be copied without constituting copyright infringement. That's not correct. This is not the way the copyright laws work. While brief phrases cannot be registered as copyright, taking essential brief phrases or portions out of a copyrighted work can be an infringement.

The old saying that "possession is nine-tenths of the law" does not work when it comes to copyright laws.

Clearance or consent

If you need to copy a prior work that is still protected under the copyright laws, it is essential that you obtain written clearance or consent from the owner of the copyright. These are really copyright licenses to use the work for a specific purpose. This is a very common practice. Of course, if you want to create an animated version of *Arthur Miller*'s play "**DEATH OF A SALESMAN**" (a bad idea, by the way), you would need permission to create a derivative work. Likewise, even though *Jane Austen*'s book "**PRIDE AND PREJUDICE**" is in the public domain, if you want to use portions of the recent motion picture based on that book in a music video, you would need permission from the owner of the copyright in the motion picture. This requirement applies even if you want to use only a simple photograph you found for an ad.

Sometimes it is easy to locate a copyright owner and other times it is not. You can start by getting in touch with his or her agent or publisher, if you can figure out who they are. In other instances, copyright owners are represented by organizations that keep track of this. If you cannot figure out who the copyright owner is or how to get in touch with him or her, and you are pretty certain the work is still protected, then don't use it!

For rights in stock photography, images, or music used in advertisements and similar purposes, the cost for a consent is usually not substantial. Getting rights to make a movie based on a book, or to create a series of posters based on a well-known painting is a different story and can be very costly.

Any consents you get will be for a very specific purpose and will probably be limited to use for a certain period of time and/or a certain number of copies.

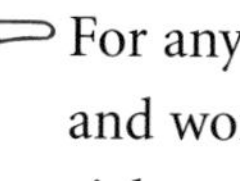

For any substantial copying or derivative works, you should hire a lawyer and work closely with him or her to make sure you are dealing with the right people and getting the consents you need. Here are some examples of clearing sources:

- **ASCAP–BMI** for public performance of music, including broadcasts (www.ascap.com and www.bmi.com).
- **HARRY FOX** if you need permission for recordings of music (www.harryfox.com).
- **GETTY IMAGES** (www.gettyimages.com) and **CORBIS** (http://pro.corbis.com) for rights in stock photography and illustrations.

DEBUNKING THE MYTH—Giving credit

Don't get the idea that being polite and giving credit to someone else's work that you have copied is a defense of copyright infringement. These are two different worlds. Giving credit is a nice thing to do and will avoid charges of plagiarism. But it is not a license to copy from the protected work. In the academic world, some copying may very well be a "fair use." This is an acceptable practice, coupled with giving credit. But this special exception does not apply to most other situations.

But don't confuse getting permission to perform or record a song with permission to use a particular recorded performance of a song by a particular artist or group.

MAKING SURE EVERYONE KNOWS WHO HAS WHAT RIGHTS

The old saying that "possession is nine-tenths of the law" does not work when it comes to copyright laws. Remember, copyrights are a bundle of rights. There are rights in the physical original such as the right to publicly display the work. There are rights to make copies. There are rights to create new works based on your existing work, such as a sequel of a book, or a recording of a song, or a poster of a painting. Giving someone one right to use your work doesn't give them all rights, unless you say otherwise. Selling an original canvas with your oil painting to someone does not give them the right to copy—for example, to make posters or calendars or stationery of the painting—although it would give them the right to display the work. And giving a magazine the right to publish your story doesn't give you the right to produce a movie with it. The same applies to software. While many CDs come with limited licenses, you need to know the terms before you start making copies. Be careful! Just because it's easy doesn't make it lawful. Here are a few categories in the creative world where it pays to know who has what rights.

Photography

Photographers own the copyrights in their photographs. They don't own the subject matter (unless they created that too), so if you want to take your own photograph of the same object, that's fine. A skilled photographer takes into account many factors when taking pictures. It is not just the push of a button that is involved. This means that, if you hire a photographer to take pictures at your wedding, you cannot just run off on your own and make as many copies as you wish. Unless this has been agreed to, it's a violation of the photographer's copyrights. You need special permission, and will likely pay an extra fee for it.

The same applies to photographs given to advertising agencies. Unless it is a work for hire or the photographer assigns all of his or her copyrights, the photograph is given for only a specific purpose. A photographer who gives a photograph to an advertising agency for an ad shot has not given up his or her rights to use it for posters or on packaging. If you are a photographer, make sure all this is sorted out when you are hired for a job so that everyone knows what they are getting out of the deal.

Music

Because of the number and different nature of ways a work of music can be used commercially, the number and complexity of rights in the musical field exceed those for other kinds of works. Each type of use has its own license, so there are different kinds of licenses for publicly performing, recording, broadcasting, and using a musical work such as a song for a movie soundtrack. You pay a certain amount of money in return for permission to perform or record a song. But don't confuse getting permission to perform or record a song with permission to use a particular recorded performance of a song by a particular artist or group. There is a separate copyright in the performer's recorded performance. Both songs and recorded performances

TIP

BE CAREFUL USING WEB LINKS

If you are building a website and are very impressed with another site and would like people to know about it, you will be tempted to provide a link to it. While linking may not be a copyright infringement, the trademark laws or laws of unfair competition might apply. We believe the wise web builder will seek permission from the web master of the site you wish to link to. He or she will probably be thrilled, but it pays to check.

of certain songs have become a rich source of music in car advertisements over the years. A lot of money changes hands, so make sure that you understand the rights you are getting before moving forward.

And by now we hope you understand why the recording industry and everyone else associated with it goes nuts when people copy or download recordings and pass them along to others without permission. A lot of time, effort, and money go into finding the right artists, selecting their good songs, getting the recordings of their performances made and engineered so they sound just right, and then getting them onto a CD or other piece of software. When you buy a CD, you are not buying just a piece of plastic, you are buying the end product of a lot of people who have put their creative expression into it. And that is all protected by copyrights. If you are a musician, a songwriter, a sound engineer, or an A&R person, you know that already!

The Internet

The Internet has become the "wild west" of the 21st century. Search engines and other specialized websites help you find all sorts of interesting, quirky, and fresh information and material. But don't be fooled. Copyright laws and other laws apply to the Internet just as they do to print. The fact that it is easy to download materials doesn't mean you can do anything you please with them. Just like recordings, small icons and logos on a website are works of art and copying them might constitute copyright infringement.

A single web page might contain numerous protected works owned by different parties—for example, text (*a literary work*), graphics (*an artistic work*), and sound files (*musical works*). The same applies to downloadable software, music, and graphics. Even databases are protected under copyright laws if their selection and arrangement of the subject matter are original. While the owners might give you a license or permission to use their copyrighted works, you should read the licenses on the web or elsewhere to make sure that your particular intended use is permitted under the license.

!

NOTE OF CAUTION!—Copyright of website designs

A problem that all too often arises is who has what rights in a website developed under contract. If you are having a website built, make sure that you obtain a copyright assignment of the website design from the website designer. If you are a web designer, make sure your client knows what rights are included in your fee.

Magazine articles

If you write an article for a magazine, you still own the copyright in the article unless there is an agreement to assign the article to the magazine publisher. By submitting the article for publication in the magazine you do, however, provide the magazine publisher with an implied nonexclusive license to publish your article, and that will include the right to republish it in subsequent issues of the same magazine.

TIP

ASSIGN COPYRIGHT IN ADVANCE

If a commercial artist is being hired for a major project, it makes sense to have him or her assign all the copyrights for a fixed fee (although the artist will want to retain the right to use and copy the work for his or her portfolio). It is in the interest of both parties to get all this straightened out before work begins.

Commercial artists

If you hire a graphic designer or other commercial artist to create an advertisement, a booklet, or even a website, make sure you know what you are getting for your money. If you are a graphic designer or commercial artist, you want to make it clear to your clients what they are getting and what they are not getting for the fee they pay. You should also negotiate with them up front what it would cost to assign all your rights to them versus what it would cost to license out certain additional rights. No one should assume that for a small payment they can use a creative work forever or for any purpose. The last thing anyone wants is to have a graphic designer sue them. And the last thing a graphic designer wants is to go after a client. That's not good business. But if it is after the fact and you discover that you need additional rights from a commercial artist, you can try to negotiate an assignment or additional license at that time. If you get an assignment, you can call it a "confirmatory assignment" and make it effective as of the date the artist created the work.

Joint ownership

This category applies across the board in all fields. When you and someone else create a work together, you both own the copyright in the joint work. You are co-owners, which means that, in the absence of an agreement, either of you is free to license the work or use it to prepare a new version subject only to a duty to share the revenues. If this causes you concern, you might want to purchase your co-owner's copyrights.

ENFORCING YOUR RIGHTS

You have spent years working on a book. Or you have spent endless days and nights working on some sort of software program. Or one of your paintings is finally accepted in an art show. Then you realize that someone else has copied your work without permission. What should you do?

As we always say, you need to go to your lawyer. If you do not have one, it is time to find one now. The lawyer will help you review the rights you have and then determine whether these rights have been infringed. If you have kept accurate records of the dates of your creations and revisions to them, and better yet, if you have registered your copyrights in your work, you should be in pretty good shape on this first step.

Next, you want to prove that your work has been copied. To do this, you first need to show that the other party had "access" to your work. Remember, in theory, it could have been created independently. If he or she really did create it independently, he or she is not an infringer. But let's get real, if the work in question is identical to yours, the court will generally conclude that there has been access and rule in your favor.

If the work in question is not identical to yours, you will first need to prove that the other party had access to your work. Then you will need to prove that the work in question is substantially similar to yours. This, like almost everything else in the law, is open to interpretation. Someone does not need

If you did not have a copyright registration when the infringement occurred, then you need to prove your actual damages and the profits of the infringer as to what they took from you.

to copy your entire work to infringe it. Taking relatively brief excerpts from a book or including one of your paintings in a photograph used in an advertisement can be an infringement. Creating a work based on your work is also an infringement of your rights.

Someone to whom you have given a right in your work for one type of use can also be an infringer if the work is used for other purposes. For example, selling someone your oil painting allows the person to display it but does not give him or her the right to produce posters of it. And creating a little icon for a single ad doesn't give someone the right to adopt it as a logo on every item sold for the next fifty years.

These are different rights that you still have. If you and your lawyer are comfortable that your copyright is valid and there has been an infringement of your rights, you can start off by sending a "cease-and-desist letter" to the infringer. If the infringer realizes a mistake has been made, he or she should be willing to settle up with you.

If you cannot settle, you can initiate a lawsuit. All lawsuits in the United States involving copyrights take place in the federal courts. The state courts do not get involved in copyright suits. But to sue, unless you are a foreign national whose work was not first published in the US, you generally need to have a copyright registration. If you prove to the court that there was an infringement, the court can order the other party to stop any further sale or distribution of the infringing articles. They can also order the destruction of the equipment such as molds and negatives used to make the copies. There are even criminal laws that could kick in. And if you were diligent and registered the copyrights in the work before your rights were infringed, you can also be reimbursed for your lawyer's fees and request automatic monetary damages up to $30,000. The court can increase these damages up to $150,000 if the infringer was really naughty. Automatic damages are known as "statutory damages," since they are provided for under the copyright statute. They are useful since it is often difficult to prove money damages. If you did not have a copyright registration when the infringement occurred, you need to prove your actual damages and the profits that the infringer took from you. This is why you should consider registering your unpublished works if others have access to them.

NOTE OF CAUTION!—Don't start counting the cash

If an infringement occurs, don't start getting excited about getting money out of the infringer. If you think someone has stolen your copyrights, remember that there might be some arguments on the other party's side. He or she might say the use is a "fair use," or that you do not have a valid copyright, or that your work is not original, or that you have given permission. So, expect a long and protracted battle unless your opponent made a clear error. Likewise, even if there has been an infringement and the court gives you a big award, you need to collect it. If the infringer is a small outfit, you may have trouble finding the company or it could go bankrupt.

This doesn't mean that you should not pursue infringers aggressively. But be realistic. All lawsuits take up a lot of time, and it could be time you would have otherwise spent creating new works or exploiting your existing works. If it is a really serious problem, you need to work carefully with your lawyers to make sure they understand your expectations.

AGREEMENTS AFFECTING YOUR COPYRIGHTS

Since your copyrights are actually a bundle of rights, like holding a bunch of sticks, always remember that you can sell or give permission to use some of these rights to one party and keep the other rights for yourself. You can sell or license these rights to someone else. This gives you a lot of flexibility with your rights. But keep good records of exactly which rights you are giving up. You should also make sure who has them, for what purpose, and for how long. If you read the preceding sections, you will appreciate why this is important.

Assignments

For works that you own as an individual, as opposed to works for hire, you can assign or transfer all or part of these rights. Assignments like this must be in writing (see Appendix 4, p.268–9 for an example). Works for hire don't need to be assigned because they are owned by the person who hires the artist or writer as soon as the work is created. But it is not uncommon for even a work-for-hire agreement to include an agreement to assign the work when it is created.

TIP

ASSIGNMENTS

This right to terminate an earlier assignment does not apply to works made for hire. In these cases, you as an individual never owned that work in the first place. If you are someone looking at buying a business where copyrights are important, you want to know whether the copyrights are works made for hire, or if they were assigned by original authors. If they are works made for hire, and you are an assignee, then you do not need to consider the ramifications about having your ownership in them being terminated at some point.

DEBUNKING THE MYTH—Another bite at the apple

You might think that, once you assign your copyrights to someone, you have given them up permanently. Or if you are getting an assignment, you might think the copyrights are yours until they expire or you sell them. That is certainly the case if you sell your home or car. But the copyright laws are different. Assignments of works by individuals on or after January 1, 1978 can be "terminated." This can be done at any time after the end of 35 years from the date of the assignment for a period of five years after that. If your work is published after the original assignment, the five-year period to terminate the assignment begins either at the end of 35 years from the date of publication or 40 years from the date of the assignment, whichever date comes first. This way, if you are a struggling artist, writer, or composer and assign your copyrights to someone else who is successful at exploiting them, you get another bite at the apple. This area of the law gets very tricky, but, if you are the creator, it can work in your favor under certain circumstances. Consult with your lawyer to make sure you don't let a deadline go by and lose your termination right. If you are the one getting the assignment, it can come as an unpleasant surprise if you have a huge investment in the work.

You can sell or license these rights to someone else. This gives you a lot of flexibility with your rights.

> **TIP**
>
> **SECURITY AGREEMENTS**
>
> You can also pledge your copyrights to a bank as collateral to get a loan. In theory, you can do this whether your work is registered or unregistered. But, as a practical matter, the bank will want them registered so that it can record the loan documents and security agreement with the Copyright Office.

Licenses

We cannot stress enough how important it is to know that you can also license out all or any portion of your copyrights. These licenses can be given on an exclusive or nonexclusive basis and they can be given out for varying lengths of time. It is like renting out rooms of your house to different people for different purposes. Exclusive licenses need to be in writing. While nonexclusive licenses do not need to be in writing, we strongly urge you to get them in writing. This way, everyone knows what they are getting. Also, the termination rights that apply to assignments also apply to long-term licenses whether exclusive or nonexclusive.

Under a license, you can collect royalties. If you do that, make sure you know what the standard in your industry is—it varies. Also, you want to know exactly how the royalties are calculated and when they are due. You must make it clear exactly which rights you are giving out and for how long. If you are giving out the right for your play to be performed, you should find out whether the performance is for a single stage performance, a television broadcast, or a movie. All these rights can be divvied up into as many pieces as you like, but you must establish when you are entering into your licenses what you are giving out. The reverse is true if you are taking a license from someone with a work protected by copyright. Make sure you know what you are getting in writing. If the license is nonexclusive, the author is allowed to give out the same rights to someone else. You obviously pay a lot more for exclusive rights, but you must understand that when doing the deal. You should be aware of two types of license: open source licenses and creative commons licenses.

► **Open source licenses** In the computer field, some software programmers make their source code freely available so that the end user can make copies of the software without payment. But, just like any other license, these licenses have their limits. The "open source license" is not on an "anything goes" basis. Any modifications or improvements or other types of derivative works require the creator of such works to distribute the modified work on the same free basis as the original open source software. It is truly a gift. The **LINUX** operating system is an example of this.

► **Creative common licenses** This category of license is similar to that of the "open source license." The exception is that the creative works are films, songs, photography, literature, and website designs. These works can be copied, distributed, displayed, or performed and derivative works can be created without charging a royalty. In return, however, credit needs to be given to the original author. Or, the work should be used on a noncommercial basis only. For example, if you are a photographer and would like to get your work out there, you can make it available for free downloads. In return, anyone displaying your photograph can do this without charge as long as you are given credit. This helps you build your reputation and in the meantime others can enjoy your works free of charge.

CHAPTER 4
trade secrets

WHAT IS A TRADE SECRET?

The next time you walk into a **DEPARTMENT STORE**, close your eyes and breathe in. Unless you have a bad cold, you will likely **SENSE** a combination of **NATURAL** and **SYNTHETIC FRAGRANCES** that can range in price from a **FEW DOLLARS** for a bottle to **THOUSANDS OF DOLLARS** for an ounce. Next, stop by

If you are a cook, a pastry chef, a winemaker, an organic farmer, a chocolate maker, a perfume designer, a computer programmer, or in a host of other businesses that rely on secret recipes, formulas, computer code, or techniques to help differentiate your products and concoctions from those of others, you will appreciate how much work it is to learn just the basics. And then there are the secrets that you finally devise that separate your creations from everything else out there. These secrets are valuable business assets and you will know one when you have it. But it is not just recipes, computer code, formulas, and techniques for foods, and fragrances that

a **CHOCOLATE SHOP** and sample some of the fantastic **CHOCOLATE CREATIONS** that are available these days. After that, try a really expensive **GLASS OF WINE** and see if it is any different from the **JUGS OF WINE** you have shared with friends. Now that you smell great and have enjoyed some **FABULOUS DESSERT** and drink, it's time to **READ** this chapter.

you might want to keep secret. Regardless of your type of business, you no doubt have a list of customers or clients who are loyal and you know exactly what they need. You have a list of suppliers and other vendors who you have come to rely on to get things done through a lot of trial and error. These people and businesses help your own business run efficiently. And if you need to get your product manufactured, you know there are lots of kinks to iron out not just to get the product made but to get it out the door and onto shelves. This sort of "know-how" can make the difference between making it and breaking it, regardless of how good your products and services are.

Examples of trade secrets

A **CHEMICAL FORMULA** for a brand-new product, **INGREDIENTS** and **COMBINATIONS** for new recipes, or any confidential **BUSINESS INFORMATION** and "**KNOW-HOW**" that can be used to help your business, help you manufacture product, and help you get it to market effectively and efficiently, which is not readily available by reading books, consulting experts, or buying off-the-shelf computer programs, falls into the category of trade secrets. Trade secrets can apply to **COMPUTER PROGRAMS**, ideas for **NEW PRODUCTS** or even **ENTERTAINMENT PROPERTIES**, **FINANCIAL INFORMATION** such as **SALES FIGURES** and **MARKETING PLANS**, and strategies like **SURVEY METHODS** used by professional pollsters.

This information is very valuable—it took a lot of time and effort to develop, it is your own private information, and it is a real asset to your business. Keeping it confidential is critical. Otherwise, your product can lose its competitive edge.

Trade secrets conjure up images of cloak-and-dagger secrecy since, typically, only a few privileged individuals should have knowledge of a particular trade secret. If everyone knows it, then it's not a secret! The famous **COCA-COLA** trade secret, the carbonated beverage flavoring "**MERCHANDISE 7X**," is said to be known to only two people in the **COCA-COLA** company. It is the poster child of trade secrets! Confidentiality, therefore, lies at the heart of trade secret protection.

If the information is disclosed, it is not confidential and no trade secret protection can be provided. So it is critical that you have practices and policies in place to keep your trade secrets confidential.

Trade secrets do not require you to jump the same hurdles as for patent or copyright protection. For instance, to qualify as a trade secret, the information does not need to be either "novel" or "be expressed in a tangible medium of expression." Prime candidates for trade secret protection are information or ideas that cannot easily or accurately be reverse-engineered.

NOTE OF CAUTION!—A source code of controversy

Trade secrets can clearly be very valuable and their misappropriation can be extremely serious. For instance, in a case involving several former employees of a chip-design software maker who left the company and formed a competitor, there was evidence that they had taken some of the software maker's most valuable source code with them, an outright theft of a trade secret. The violations were so serious that criminal charges were brought. In addition, hundreds of millions of dollars had to be paid.

Trade secret is the vehicle often used to protect valuable new technology. The protection may only last for a limited period of time—for example, until patent applications for that technology are filed and published or issued. Or, when there is no commercial reason for the trade secret ever to be disclosed, it may last indefinitely.

DEFINITION—Reverse engineering

Reverse engineering is when your competitors, through careful analysis, are able to figure out the different unique components, their composition, relative quantities, and assembly of your product or service.

DO I HAVE A TRADE SECRET OR AN INVENTION TO PATENT?

If your particular trade secret is also new, novel, and nonobvious, you may want to consider whether it would be advantageous to protect it as a patent.

The key here is whether your creation can be readily reverse-engineered or is likely to be figured out by your competition. For instance, can your competitors, through careful analysis, figure out your actual techniques, the amount and proportions of components and ingredients, and the exact steps of your process? If your creation is really just ahead of the curve and through independent research your competitors are likely eventually to figure it out, you should seriously consider filing a patent application.

Otherwise, you may be better off keeping your creation as a trade secret. As a trade secret, your creation is not limited to the patent term of 20 years and may continue as long as it is kept confidential. For example, **COCA-COLA**'s secret formula has been a trade secret for over 100 years and is still going strong. Also, the long-term cost involved with trade secret protection is often substantially less than obtaining and maintaining a group of patents.

Trade secret protection and patent or copyright protection often complement each other. This is where you may wish to consider bundling your various rights. Careful strategy may provide you with substantially more layers of protection than if you relied on one set of rights only.

If you get into a dispute with someone who steals your trade secret, you will want to be able to prove what you had at a given time and that you kept it secret.

For instance, you may wish to keep some specific part of your developments for trade secret protection. The remainder of your developments, which can be reverse-engineered or developed by others, could possibly be protected as a patent. Keep in mind though, that if you choose to file a patent application, you need to play by the government's rules. This includes disclosing the best way for people to make or use the invention in your claims. You need to review this strategy carefully with your intellectual property professional. Furthermore, some aspects may also qualify for copyright protection. For example, certain aspects of computer programs may qualify not only for trade secret protection but also for patent and copyright protection. Always consider all possible scenarios in order to secure maximum coverage for protection. These three types of protection are not mutually exclusive.

HOW DO I GET TRADE SECRET PROTECTION?

Trade secrets don't get registered, since they are by definition your own secret materials. Most of the law in this area for enforcing your rights is determined by the states. Unlike utility patents, trademarks, and copyrights, there is no official federal trade secret law. But in general, if you get into a dispute with someone who steals your trade secret, you will want to be able to prove what you had at a given time and that you kept it secret. The following sections give you some pointers on how to do that. But you should also consult your lawyer to make sure all the steps you take will work for your business.

Unlike utility patents, trademarks, and copyrights, there is no official federal trade secret law.

Keep it secret

You must be able to show that you are keeping your trade secrets confidential. If you ever get into a dispute, you will need to prove this point convincingly. Keep good records of the steps you have taken to ensure the physical protection of your trade secret and restriction of access to it. For instance, is your secret formula kept in a vault for safekeeping? Always remember that access to the information should be highly restricted. Having a detailed business policy for handling your trade secrets is a good start. This applies to chefs too! Your policy should include the following steps:

- When trade secret information is shared, make sure that correspondence is at all times clearly marked as "Confidential".
- Don't leave confidential materials or apparatus out in the open.
- Don't permit public tours of your manufacturing facilities or other areas where outsiders may have access to your trade secrets.
- Don't give your own employees access to your trade secrets unless they need them to perform their tasks.
- Make sure all employees and other essential people like consultants sign a confidentiality agreement.
- Keep written records of who visits, when they visit, and what the purpose of their visit is. Larger companies make all visitors wear badges so everyone knows who a visitor is. Whatever you do, if trade secrets are involved in your business, you need to control the whereabouts of people on your premises and the handling of confidential materials to show that you have a policy of keeping your information secret.

You must be able to show that any disclosures of your trade secret have been under an obligation of confidence.

Confidentiality agreements

You must be able to show that any disclosures of your trade secret have been under an obligation of confidence. By far the best way to prove this point is to request that everyone to whom you disclose your trade secret sign a "confidentiality letter" or "nondisclosure agreement or "NDA." This includes all your employees and consultants, even if you do not think they will have access to your trade secrets. It is no wonder that people have been handing out one-page NDAs like business cards at commercial gatherings ever since the dotcom revolution. For an example of a typical NDA see Appendix 4, pp.270–1. Remember: the smaller the circle of people who are in the know, the better—not only for commercial security, but also for legal reasons. Even a chef's assistant should sign an NDA—don't be shy about this, it's your creation at stake.

Remember, "an ounce of prevention is worth a pound of cure."

NOTE OF CAUTION!—Resistance to sign NDAs

However, we warn you that you will often encounter resistance from big corporations, which may not be willing to sign an NDA when you approach them with a proposal for commercial cooperation. Their reason, sometimes justifiably so, is that they may be working on a similar project. In these circumstances, some people disclose the end result of their trade secret without revealing the confidential know-how. This should work most of the time, but sometimes it is impossible to reveal the end result without the trade secret itself. In these cases, when talking to a corporate gorilla, you will have to either assume the risk in order to profit from a possible business opportunity or consider filing a patent application where feasible. You may also want to give thought to slapping a copyright notice on your original written materials before handing them over. Copyright is not the ideal vehicle to deal with protecting confidential information, but the notice may have a deterrent effect. Remember, "an ounce of prevention is worth a pound of cure."

Keep good records

You want a good written record of what you have at any given time. Since there is no official filing, it is your records and how they are kept that will help establish what information you had at any given time. And keep the records in a safe place where only the most trusted people, who have signed confidentiality agreements, have access to them.

Remember: the smaller the circle of people who are in the know, the better—not only for commercial security, but also for legal reasons.

WHEN DOES SOMEONE VIOLATE MY TRADE SECRET?

If someone who is under an obligation of confidence discloses your trade secret, it is a violation of your rights and may rise to the level of an outright criminal act. Moreover, anyone else who acquires this information with knowledge of its misappropriation can be liable to you. Likewise, someone who breaks into your space can violate your rights. But proving what you had and showing you kept it a secret can be a tough job. This type of case can also involve copyright infringement. You should consult an intellectual property professional at this juncture who will help sort out what rights you have and whether they have been stolen or violated. Then they can advise you on litigation strategy or other action to take.

CHAPTER 5 utility patents

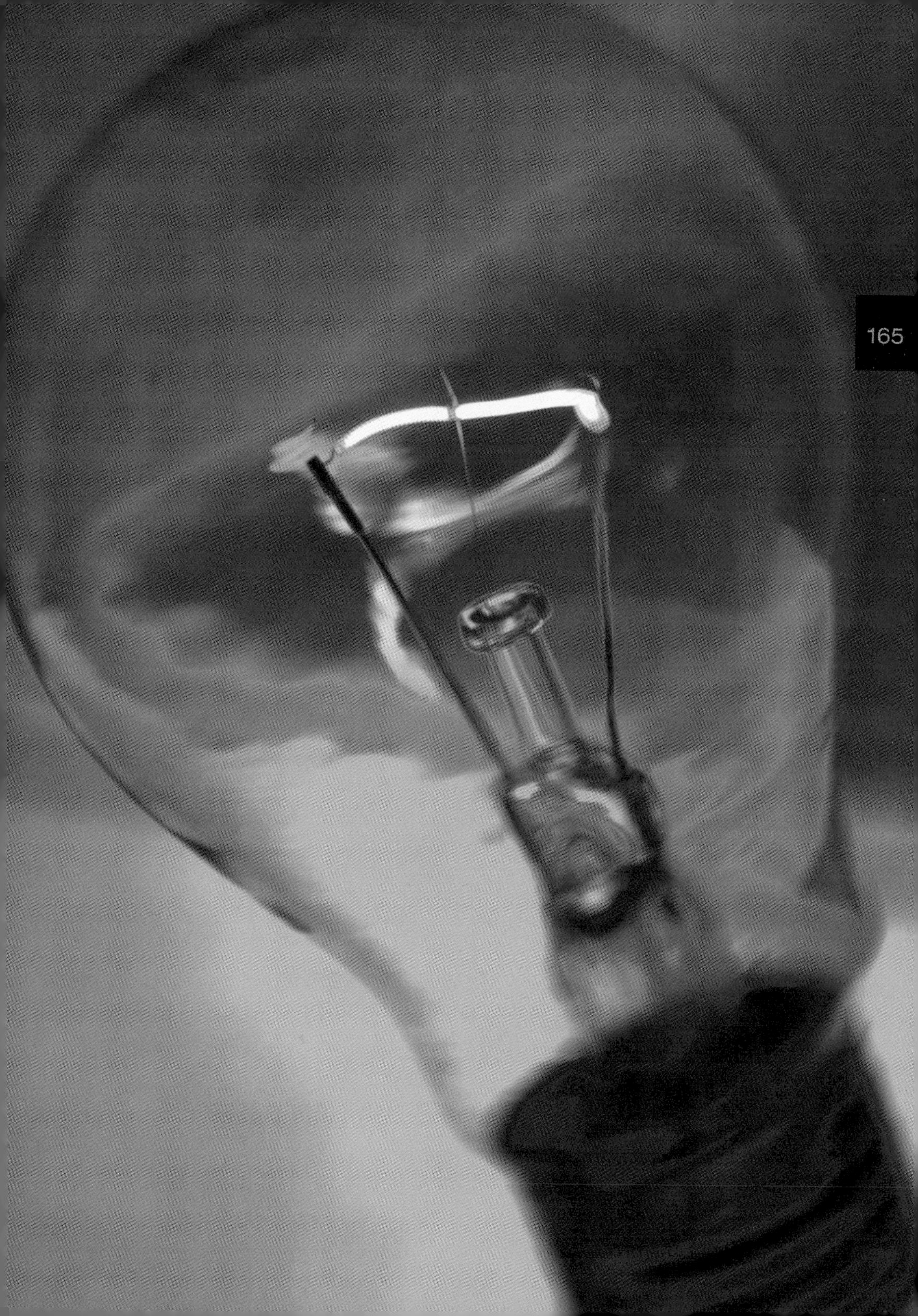

WHAT IS A PATENT?

When you turn off the light at night and your head hits the pillow, the last thing running through your mind (at least we hope) is whether your **PILLOW** has any **PATENTED FEATURES**.

The truth is, wherever you are sitting as you read this paragraph, if you look around you, it is likely that within a few feet are items that are patented or are made under a patented process. This includes the chair you are sitting on and the method of weaving the carpeting underneath you. And if you are on the beach reading this, there is probably a patent on your beach chair, or the material your beach umbrella is made of, or the suntan lotion you just put on. And if you are at your desk, there are patents covering your computer hardware, and even aspects of your computer software. And your telephone—the satellite system or land lines, as well as all the equipment, are or have been patented.

You don't need a laboratory where everyone runs around in white coats speaking in hushed tones for inventions to be created. And you don't need

The same goes for the **MATTRESS**—what about patents on that? Who invented the **PILL** you just took? What kinds of patents are protecting the **LAMP** on your **NIGHTSTAND**? And hey, don't forget the **LIGHT BULB**!

to have bulging eyes and a hunched back to be an inventor. Such locations and people may very well create great inventions. But it is just as likely that your best friend's mother, or the person sitting next to you when you go to renew your driver's license, and yes, even you, have as much of an opportunity to invent something as anyone else. Will your invention change the course of humankind? If it does, be sure to tell everyone that you read this book. But your invention can be just as worthy of patent protection if it is an innovation, as long as it is not an obvious improvement on what is already out there, and is useful to others.

The people who conceive of inventions are all influenced by everything that came before. They are smart enough, reasonable enough, and motivated enough to take the time and effort to recognize a new idea. But we all come

In the United States, there are three types of patents. They are known as utility patents, design patents, and plant patents.

up with an idea or a dream. It is turning these dreams and ideas into actual working things or processes, that is what invention is about. This is what separates inventors from the rest of the creative pack. Inventors are skilled enough and talented enough to build something or develop a process with their dreams and ideas.

It is not just earth-moving innovations that affect everyone's lives that can be protected under the patent laws. There is always room for improvement. You can improve on existing things and build on them, making them better, or easier, or cheaper, or more efficient to use. You can also expand on basic inventions. After all, the light bulb is great, but laser light technology has brought about both incredible medical advances and lots of entertaining light shows.

Anyone who has an original idea **AND** takes the time and effort to develop it into something usable is an inventor. And the people who improve and build on these foundations are also inventors. If you are one of them, you need to know how to protect your inventions, how to reap the rewards of your efforts, and how you can build on these to get a better product into the marketplace. It is not just genius but perseverance that counts.

OK, so inventions are protected by patents. But what is a patent? How do I get one? What type of rights does it protect?

TYPES OF PATENTS

In the United States, there are three types of patents: utility patents, design patents, and plant patents. Utility patents are by far the "big daddies" in the world of patents. Usually, when someone is talking about a "patent," they are referring to utility patents. That's what we will do here too.

DEFINITION—Patents

A patent protects new or improved apparatus or processes as well as chemical or biological combinations. This protection can cover functional or technical features. Design patents protect ornamental features of manufactured products and are discussed in the next chapter. Plant patents are one of those esoteric areas. They protect innovations or discoveries of asexually reproduced hybrid plants and newly found seedlings. That's all we will say about that.

Patents can cover any technology or field, but they are generally characterized as electrical, biotechnology, software, business methods, chemical, and mechanical. Patents cover hundreds of thousands of products you see in day-to-day life, ranging from vitamin pills to flat-screen televisions, to the windshield wipers on your car. They also cover satellite systems, methods for downloading music, and endless types of plastics. And artificial implants? Patented. The methods for implanting them? Patented. You get the idea. It's a miracle that most of us get through the day without getting a demand letter from someone saying we have infringed their patent!

Patents can also cover improvements that are made to patented products and processes over time.

Examples of inventions protected by patents

The **TELEPHONE**, the **LIGHT BULB**, **AMAZON**'s **ONE-CLICK CHECKOUT** system, most **PHARMACEUTICALS**, including the industrial production of **PENICILLIN**, **WINDOWS** software, the **BALLPOINT**, **ARTIFICIAL HEARTS**, a **COMPUTER MOUSE**, are all examples of technology protected by past or current patents.

What do patents protect?

Patents can protect a basic product and the way it is made, as well as specific features for that product or process. It does not give you a monopoly, but with your patent you can seek to exclude others from making, using, offering to sell, exporting, importing, selling, or marketing your invention.

Most patents last for 20 years from the filing date. But there are maintenance fees to pay along the way. Patents cannot be renewed. Once your patent expires, your invention is freely available. So you want to get the most out of your patent while it is alive.

It is not at all uncommon for a company to own many patents that cover what you might think is one product. Patents can also cover improvements that are made to patented products and processes over time. So companies can also have a series of patents that extend over a long period of time. By introducing improvements, consumers benefit. So do the patent owners, since the fruits of their development activities in respect of existing products

If done right, a patent will prevent your competition from making or using or selling whatever is covered by the claims of the patent. It gives you a competitive edge so that you can develop a group of loyal customers who will want your product even after the patent expires. And if the demand for your patented product is greater than you can handle, you can license it to others and get royalties.

and processes lead to new patent protection. Some people call this built-in obsolescence. We call it development and progress. If done right, a patent will prevent your competition from making or using or selling whatever is covered by the claims of the patent. It gives you a competitive edge so that you can develop a group of loyal customers who will want your product even after the patent expires. And if the demand for your patented product is greater than you can handle, you can license it to others and get royalties.

It is in a patent that your hard labors are defined.

What's inside a patent?

When you first see a patent it can be off-putting. Visit www.edisontoipod.com for an example. They often have a technical drawing on the cover and a description of the invention. This is called the "abstract." But don't judge a patent by its cover. You need to open it and start reading it to really understand what is covered. A patent has several sections and varies in length depending upon the subject. It describes the invention and its background.

Patents are generally accompanied by detailed drawings, which must clearly and accurately show all aspects of the invention. Background to the area of the invention is provided so that you can understand how this invention fits into the real world. And an explanation is provided as to why and what the inventor has come up with and how this invention is new and sufficiently different from everything else out there to deserve a patent. This is followed by a detailed description explaining the various elements shown in the drawings and how they come together and work as a whole. At the end of all this is a series of numbered paragraphs that carefully and consistently define exactly what your invention is. These are known as the "claims," and provide the legal definition of your invention. They stake out the land that your invention covers. These can be difficult to read, as they are very detailed. But, like Shakespeare, after a few minutes of reading claims, you can start to follow them. They should make a lot of sense, even if they are technical. But it is everything as a whole in your patent that will define your invention and give meaning to what you claim.

What are the basic requirements for a patent?

You cannot get a patent on any old thing. But show us a new thing and we bet there is an invention involved, maybe even several inventions. Just make certain that it is useful, new, and nonobvious compared to everything else that was out there before your invention.

Useful Your invention must accomplish a practical application and produce a useful, concrete, and tangible result. This requirement is generally not a problem. Abstract ideas or mathematical formulas alone, as opposed to their applications, cannot be the subject of a patent because they are not "useful." But determining what is useful and not useful can change as our culture evolves. Prior to the "**DOTCOM**" era, it was generally believed that methods of doing business could not be patented. However, once the Internet got out of the scientific and academic community and into the hands of entrepreneurs and consumers, our culture changed. So did the way of doing business. And the Patent and Trademark Office started to change its tune and started to

Just make certain that what you want to patent is useful, new, and nonobvious compared to everything else that was out there before your invention.

Prior art includes publications that were available anywhere in the world, and any public uses within the United States.

issue patents on these innovations. Many of these patents, such as **AMAZON**'s **ONE-CLICK CHECKOUT**, have come to be known as "business method patents." In some cases, the range of protection has become so extreme that critics refer to this change as the "silliness standard." The Patent and Trademark Office is looking at this carefully. But this type of protection lives on and reflects the new era in which we live.

New One of the most important requirements for getting a patent is that your invention needs to be new, or in patent talk, "novel," compared to all the information and materials that were out there before your invention.

DEFINITION—Prior art
All earlier materials and prior information are known as prior art. Prior art includes other people's stuff, as well as your own stuff. It includes publications that were available anywhere in the world, and any public uses within the United States.

If the date of a publication or use was before the date of your invention or more than one year prior to the date you file your US patent application, and if this prior disclosure includes everything you claim as your invention, you are out of luck. You cannot claim that invention in a patent. This is called "anticipation." Prior art also includes patents filed before the date of your invention. And it's not just the claims of prior patents that are considered "prior art," it's everything the prior patent shows or describes

TIP

CLAIMING NOVELTY

You can be denied a patent for lacking novelty only if one prior art reference discloses each and every feature of your claimed invention. When you stake out your claims in your patent application, you need to make certain that your claims do not cover one reference that is already out there.

and all other publications. And here is some stuff that you do that can be prior art. If you intend to sell or offer for sale a product or process that you want to claim as your invention, you should file a patent application before you do. If you want to protect your invention only in the US, you have a one-year "grace period" to get on file, from the point you make any sales or offers for sale. If you fail to do that, you are out of the running for a patent. This also applies to any public uses you make in the US of your new product or process. You also have only one year to file after you show or describe your invention in a publication anywhere in the world (including foreign patents).

This is one moment of your life where it does not pay to have a big mouth or be a show-off—save that for later.

!

NOTE OF CAUTION!—The 10-ton weight hanging over your head

It's vital to understand that your own activities in relation to your invention can cause you to lose the ability to patent it. Imagine a 10-ton weight hanging over your invention with a big clock on it. Once you take one of the above steps, the clock starts to tick. Once the clock reaches one year, the 10-ton weight comes crashing down. If you have not filed a patent application covering your invention, your US rights in it are crushed to smithereens. Don't let that happen to your inventions.

TIP

SEARCHES

Prior art patents in the United States and elsewhere in the world are relatively easy to retrieve these days. Since there might be prior art that can prevent you from getting a patent because your claimed invention is not new, doing a search of patents and other materials at least in the United States, and preferably worldwide, before filing your application can be very useful. This way, you can word the claims to skirt around this stuff. Patent searches are described in more detail below.

The key here is that you have only one year to test the US market before filing a patent application and no "grace period" if protection in foreign countries is of interest. So you need to keep track of your activities and make sure that you are on file in the United States Patent and Trademark Office within one year. This is one moment of your life where it does not pay to have a big mouth or be a show-off—save that for later. It's best to file first and disclose later!

This requirement that your claimed invention be "new" can sound scary and overwhelming at first. But remember, you are in control of your own actions, so make sure you get on file within a year, if not sooner, after you do something that could cause you to lose your rights. Remember, if you want protection in foreign countries, you have to file before you make any disclosures. Otherwise, you can lose out on getting patent protection outside the United States. It's a tough world out there.

Nonobvious Your claimed invention must also not be an obvious modification of what was already out there before your invention. An Examiner might cite combinations of prior patents or other articles or even ads for earlier products you sold, and refuse to accept your claimed invention. He or she will tell you that it is obvious to come up with what you are claiming based upon what is already out there. This is often a difficult hurdle to overcome with Examiners in the Patent and Trademark Office during the patent application process. You will need to convince the

You will need to convince the Examiner that your claimed invention provides some benefit or advantage that goes beyond a mere combination of what has already been done.

Examiner that your claimed invention provides some benefit or advantage that goes beyond a mere combination of what has already been done. Oftentimes, you can tweak the wording of your claims. In doing so, you are cutting back on what you stake out, so do it carefully.

This requirement of nonobviousness is tougher to deal with than novelty. With novelty, it is pretty clear whether what you claim is already out there. But what is obvious to an Examiner may not be so obvious to you and others. The patent law recognizes this and uses a yardstick to help the Examiner make this decision.

LEGAL TEST—Is it "obvious"?

The question is whether or not the claimed invention is obvious to a person skilled in the relevant field. Even this requires you to define a fictional person and can be a little off-putting. This "dude" is neither a leading expert in your field nor someone who tinkers around in their basement on weekends. Instead, they are considered someone who is well rounded, with wide knowledge of the field where you are staking your claims.

In the US, you do not need a lawyer to file a patent application. You can file on your own. But you can also give yourself open heart surgery. DON'T DO IT.

WHAT SHOULD I DO TO PROTECT MY INVENTION?

There is one simple answer to this question: make your way to a patent professional as soon as you have a concept that you can turn into a new or improved product, or come up with a new or improved process for doing something.

Unlike trademarks, where you can rely on unregistered common law rights and copyrights that exist from the moment of creation, your inventions are not protected until a patent is issued. Because of this, many inventors are still figuring out exactly what their invention is when they go to their patent professional.

If your patent professional thinks you have something that you can protect, make sure he or she files a patent application in the Patent and Trademark Office that covers your "invention" as soon as possible. And let the professional know if the clock with the 10-ton weight is ticking and when it will go off.

!

NOTE OF CAUTION!—Don't be your own worst enemy

Remember, you can be your invention's worst enemy. If you wait to file for more than a year after you start to sell or publicize or use your invention, your own activities can prevent you from getting a patent.

If your patent professional thinks you have something that you can protect, make sure he or she files a patent application in the Patent and Trademark Office that covers your "invention" as soon as possible.

In the US, you do not need a lawyer to file a patent application. You can file on your own. But you can also give yourself open heart surgery. Don't do it. A patent agent who is registered to practice before the United States Patent and Trademark Office can also file your patent application. But patent agents are not lawyers and they are not allowed to give you legal advice. If you decide to use a lawyer, you need a "patent lawyer." Just like an agent, patent lawyers also have to be registered to file or work on patent applications in the United States Patent and Trademark Office.

There are about 7,500 active patent agents and 25,000 active patent attorneys licensed to file patent applications. Selecting a good patent lawyer or agent to prepare your application and persuade the Examiner of its merits is a big plus and can significantly ease your way through the application process. We will refer to the lawyer or agent here as your "patent professional." You can access a list of patent professionals at http://www.uspto.gov/web/offices/dcom/olia/oed/roster/region/index.html.

The official filing date of your patent is also a key to getting protection around the world. If you file your foreign applications within one year after your full application with claims is on file, you can extend it internationally. Your US filing date will also begin the 20-year period during which any patent you ultimately receive in the US will be alive.

Before you grab your coat and run out the door to see a patent professional, pause for at least a moment to think about whether it makes more sense to keep your invention as a trade secret (see more about this at Chapter 4, pp.152–163). While a patent will give you the right to exclude the competition from using your invention, that right does not last forever. In most cases, the commercial importance of your invention will be outdated before the patent expires. But if your invention is of the type that people will want to use or exploit for decades, and it is of the type that people cannot reverse engineer (in other words, they probably won't find out about it from anyone but you), it might make sense to treat it as a trade secret. Think of the **COCA-COLA** formula (see p.156). It was never patented, and only a few people know exactly what goes into that carbonated beverage to give it that famous **COCA-COLA** taste. You should also carefully analyze whether your invention overlaps or may be "bundled" with other forms of intellectual property. For more information on bundling your intellectual property, see pp.234–241.

In our view, one of the best practices to help you keep out of this type of trouble is to keep good notes of what you are doing each day.

Even if you have not made any prior disclosures to start a one-year clock, the sooner you file, the better. If you are in a field where developments are coming fast and furious, your delay could let other people file before you and their patents might be cited against you. Before you get to your patent professional, here are some good practices to follow.

Keep good records

There are two reasons that you should keep good records of your day-to-day activities when developing an idea into something that you would like to claim as an invention. Under US law, "invention" is made up of two steps, conception and reduction to practice. If the date of "invention" is contested, the steps need to be proved. Once on file, there might be certain types of "prior art" cited against you because it is dated before your filing date. But if the actual date of your invention is before your filing date, you might be able to submit evidence to prove that the prior art is not "prior art".

!

NOTE OF CAUTION!—Accusations of infringements from others

Another reason for keeping good records is more defensive in nature. You may be working on certain developments that you have decided not to patent, but which you start to use in your business. If someone else gets a patent on these developments, they might sue you for infringement. If you can prove through your documentation that you were already doing what is in the plaintiff's claims, in many cases your own activities will be viewed as prior art and you are off the hook.

In our view, one of the best ways to stay out of this type of trouble is to keep good notes of what you are doing each day. You should sign and date every page of your notes at the end of each work session. Then have someone sign and date the page as a witness to what is on that page. You should do this as close as possible to the date that you make the entries. Otherwise you have no one to back you up as to the date on which you were already doing something. But here is where you need to be extra-cautious. You should separately enter into a written agreement with your witnesses to keep any information disclosed to them in confidence. They should confirm that they are looking at this information only to witness to the date they are signing off on these pages. Otherwise you might start the one-year clock with the 10-ton weight.

Don't disclose your invention!

We already told you about this. But some things deserve a repeat performance and this is one of them. And we've added a twist of lime in case you're getting bored. This rule sounds obvious and easy to comply with. But it's not. If you are a creative type, you are also probably pretty enthused about what you come up with. Inevitably you will be tempted to tell your friends or family or co-workers about what you are working on. But be aware, very aware. We've already warned about the one-year rule as to certain offers for sale or public uses or publications you might make. But in certain parts of the world you do not get any leeway.

So even if you can file an application in the United States, if your invention has a global market, you might lose out on the ability to get protection outside the United States. Here are a few examples.

▶ Talking to acquaintances at a meeting or cocktail party (we told you we'd add some lime) might not immediately threaten your ability to file in the United States, but it could definitely cut off your ability to file elsewhere.

▶ Publishing a short discussion of your invention on your website or in a chat room or a not-widely-circulated newsletter could throw off your international rights and also start the one-year clock running in the United States. The same applies if you give a lecture at an academic conference or describe your invention to a potential customer.

DEFINITION—Nondisclosure agreements

There will inevitably be people you want or need to discuss your invention with. If you absolutely need to discuss your invention with anyone before your patent application is filed (and we really mean any living human being), you should first have that person sign a confidentiality or nondisclosure agreement (NDA) (see Appendix 4, pp.270–271, for an example). All the rules on confidentiality described in the trade secrets chapter apply here. In fact, there is an overlap between trade secret and patent protection in this situation. Your invention can be protected from stealing as a trade secret until such time as your patent application is filed. If you have a business reason to keep your invention secret, even after filing, you can continue to follow these rules until your application is published by the Patent and Trademark Office.

A full-fledged patent application can be expensive to file and very time-consuming to prepare.

You should follow the suggestions on confidentiality outlined in the trade secrets chapter at least until your application is filed. But even if the discussions are confidential, if they amount to an offer to sell what you plan to claim, the one-year clock will start ticking if you are not on file. Discussions that you have with your patent attorney are privileged and also confidential, so these can be made without a confidentiality agreement. In fact, your patent attorney can provide you with a suitably worded NDA.

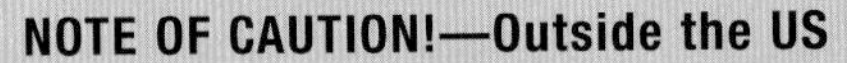

NOTE OF CAUTION!—Outside the US

If you disclose your invention before you get on file, you will normally still have one year to file your US patent application with the US Patent and Trademark Office before that disclosure can come back to haunt you. However, almost anywhere else in the world life is not so forgiving—even if your disclosure is made one day before you file.

Don't offer to sell your invention before you file!

And here's a repeat performance. Even if you do not disclose any details of your invention, by making an offer to sell the actual invention, or even a product that incorporates your invention, you risk starting the one-year clock and you will need to get on file in the United States. If there is genuine commercial interest by someone in your invention, you may not be able to hold back. But once you make that offer, remember, you will have only a limited period of time to get your application on file. And your offer might also cause you to lose the ability to get foreign patent protection.

TIP

PATENTABILITY SEARCHES

Patentability searches can be very helpful in identifying a wealth of technical information. For example, you may be able to identify subcomponents that have already been invented and could support the development of your invention so that you don't have to reinvent the same technology. In addition, it is useful to find out what your competitors have been up to!

Provisional applications

A full-fledged patent application can be expensive to file and very time-consuming to prepare. Therefore, in recent years the law has changed so that you can file a "provisional" patent application. This is essentially an informal application that does not require claims. Instead, a full description of the invention is provided. You then have one year from the date you file the provisional application to turn it into a more complete application with a full set of claims.

If, during that year, you decide that what you came up with does not make any sense, or has no commercial value, you can abandon it. If that happens, what you disclose will stay confidential. If you end up filing a full application with claims, the filing date for anything you disclosed in the provisional application is the date the Examiner and courts will look at when they consider whether prior materials should prevent you from having a patent. If you do not have a patent attorney or patent agent by this time, you should get one to file your provisional application.

Patentability searches

It is a good idea to have a patentability search conducted before filing your full patent application, that will have claims. If the search locates prior patents on other materials that show that your invention is not new or is an obvious improvement, you will have saved a lot of time and hassle pursuing a patent. The results can also help you and your lawyer decide what to focus

TIP

PATENT DATABASES

In a patentability search you essentially compare what you plan to claim as your invention against the existing prior art found during the search. Patent searchers will typically conduct such searches by using two sources: electronic patent databases and trade or technical journals and databases in your field.

The two main patent databases that are freely accessible to the public are the US Patent and Trademark Office database (http://www.uspto.gov/patft/index.html)—which includes 7 million patent files—and the Esp@cenet database administered by the European Patent Office (http://ep.espacenet.com)—which houses 50 million patent documents. If you have limited funds, you can try to conduct your own search using the above sources. However, searching requires considerable care and patience, and the above databases (in particular Esp@cenet) have, by design, significant limitations so that your ability to search prior patents in detail is restricted. As a result, your own search will not likely be as comprehensive as one done on your behalf by a patent professional.

on in your application. So even if the search dredges up stuff that looks like your invention, there might still be hope. Often you can word your claims so that what you claim is new and not obvious compared to the search results.

!

NOTE OF CAUTION!—It's not exact

The downside of doing searches is that they can be very expensive and time-consuming. Also, someone might file an application before you while you are still searching. Searching is also an inexact art, so there is no guarantee that a search will uncover all prior materials that are most relevant to your invention. On the whole though, long experience dictates that the prudent route in most cases is to consider a patentability search before filing the full application.

Apart from patentability searches, if you or your inventor are working on the cutting edge in your field, you or your inventor will be aware of "what is out there" already. Give this information to your patent professional. It can be put to good use in drafting the patent application.

Commercial viability

Patents are by far the most expensive form of intellectual property. Even though your patent lawyer will be a highly skilled professional, he or she still needs to spend the time to get to know and understand your field and your particular invention. Thus, getting a patent can be expensive and can take a couple of years. Patent litigation is one of the most costly forms of litigation. For instance, big pharmaceutical and chemical companies often continue court battles after the patents have expired in anticipation of huge damage awards running into billions of dollars. So it makes real sense for you at the start of your patent process to take a hard look at the commercial prospects of your invention to fully assess the risk/reward ratio. You may wish to consider the following factors in weighing up the pros and cons:

- Is there a need for your invention in the market?
- What does the analysis of competitors and their products reveal?
- The timing of the launch of your invention into the market should be assessed—inventions often fail because they are introduced to the market too soon or too late.
- How cost-effective will your invention be to produce?
- Are there potential strategic investors who will fund the development of your invention?

These and similar commercial issues should be seriously reviewed before taking the first step to making the financial outlay to support the filing of a patent application. You should be aware that invention promotion

A simple word, or the way the words in your claim are phrased, could make the difference between whether or not a competitor's product infringes your patent.

companies may sometimes be willing to undertake a market feasibility study of your invention. Be careful, as some of them may be less than reputable. Or they might not devote the time your invention deserves and get a very narrow patent that doesn't cover what you think it does.

FILING A FULL PATENT APPLICATION WITH CLAIMS

So here you are. All primed and prepped. Ready to present the invention to the US government to get the big stamp and ribbon on your invention and ready to enter the Inventors' Hall of Fame. Not! Sorry, your work has only just begun.

As we previously explained, in order to file your patent application you have to provide a detailed description, which is essentially a written explanation of your invention. This is known as the "specification." Generally, you also need drawings. The Patent and Trademark Office has very detailed rules for drawings. Shop drawings or even technical blueprints will not work. You should use a good patent draftsperson who can prepare the drawings under the guidance of your patent professional.

The specification and drawings are followed by the claims. We can't overemphasize how important your claims will be. They are at the heart of any patent, since they define the scope of protection and exclusivity provided by the patent. Your claims are legal statements typically captured in single sentences. They pinpoint and define your invention by describing its distinctive technical features. It is very important that the claims be extremely carefully written—they must be crystal clear, concise, and consistent with one another. In a dispute involving your patent, the meaning of one single word in the claims could be the crucial factor in determining whether your patent is valid or invalid due to some prior patent out there or a prior disclosure you made.

The clarity of your claims is equally important to put everyone else on notice as to precisely what they may or may not do to avoid infringing your patent. A simple word, or the way the words in your claim are phrased, could make the difference between whether or not a competitor's product infringes your patent.

We can't overemphasize how important your claims will be.

Usually the claims start with a broad claim that casts the inventor's net of protection as wide as possible while still avoiding all known prior art. The broad claim is normally followed by narrower claims, which should lead to stronger claims against any prior art references. These claims recite additional features of your invention.

TIP

ERR ON THE SIDE OF CAUTION

Your patent attorney can advise you whether information is material and needs to be disclosed. It is always safe to err on the side of disclosing something. You might not even know whether something is prior art, but it is better to disclose it. Then, if the Examiner cites it against you, you can look at whether it really is prior art. Prior art that is the same or interchangeable with other prior art does not all need to be disclosed. So, if you have read 15 articles, all of which say the same thing, you need to disclose only one of them—preferably the one that discloses the most information. If in doubt about whether you should disclose something, disclose it. This is one time that it is better to spill your guts.

Invention disclosure

You know your invention in detail. Your lawyer does not. As a practical matter, in getting ready to meet with your patent lawyer or agent, the best way to proceed is to first prepare a written description of what your invention is. This description does not need to be overly detailed. Your lawyer will work with you on that. But it should communicate the full nature and scope of your invention.

Keep in mind that detailed drawings are almost always required, so we encourage you to use those too. This disclosure should also include copies of materials you know about that inspired you or that you used to develop your invention. Your lawyer will then use your invention disclosure to translate your description of the invention into "patentese." This will include a small summary or "abstract" that will show up at the beginning of your patent.

Obligation to describe the best way your invention works

The patent specification has to be detailed enough so that someone "skilled in the art" in your field *(our fictional dude with an average level of skill, see p.177)* should be able to build or use the invention.

TIP

GETTING THE SPECIFICATION RIGHT

Even though you are using a patent professional, keep a very close watch on the drafting of the patent specification. In our experience, the best patent specifications are usually a close joint effort between the inventor and the patent professional. It is not uncommon during this process for the professional to help you understand what it is about your invention that sets it apart from your competitors' products.

DEBUNKING THE MYTH—Spill your guts

Inventors sometimes think they can file for a patent while still keeping key details of their invention in their back pocket in an attempt to prevent anyone else from ever being able to implement their invention as well as them. Don't be piggish! The government is giving you a patent in exchange for complete disclosure of your invention. If you fail to describe what you personally believe to be the "best mode" for implementing your invention, your patent will be invalid.

Obligation to describe material prior art

At the time of filing your application and also during the entire process until the patent issues, you are under a continuing obligation and duty to disclose to the Patent and Trademark Office all prior art that you are aware of, or learn of through your searches, or that you become aware of while your application is pending, which is "material" to getting your patent. Your patent attorney will explain this obligation to you further. If you fail to maintain this standard, the resulting patent you obtain could be unenforceable. This is so even if a piece of prior art that you do not disclose would not ruin your chances of getting a patent. We cannot stress enough how important it is to let your patent attorney know about prior articles you have read, devices you have seen, people you have spoken with, and any other activities that influenced you during the creative process when you were coming up with your invention.

If in doubt about whether you should disclose something, disclose it. This is one time that it is better to spill your guts.

Inventorship and ownership

Unlike a trademark, which can be owned by a corporation if the corporation is using the trademark, and unlike a copyright, which can also be owned by a corporation if it is a work for hire, inventions are created by individual people, not corporations. This means that, when you file your patent application, you have to identify the specific individual or individuals who conceived the specific invention and stake this out in the claims of your patent application.

We have all come up with an idea that we think about "patenting." But, as you know by now, you can't patent an idea. You need to get the good ideas out of your head and into a working prototype or description. This requires a lot of sweat and time. It separates the dreamers from the inventors! However, it is not uncommon for two people working together to contribute to what ends up being one claimed invention. This can happen among friends. Or it can happen in a research and development environment, where people are working together to accomplish common goals.

Whether your invention was created by you in your garage, on a napkin over lunch with a friend, or in a sealed laboratory where everyone wears white coats and hairnets, you need to get the name or the names of the inventors correct when you file your application. Don't insist on naming yourself as an inventor because of pride or ego, or improved job opportunities. And don't let anyone else do that to you if you are the true inventor or one of the inventors. If the wrong names are provided, or if an incomplete list of inventors is identified in the patent application, the entire patent could be jeopardized because of an unintentional false statement.

Besides, the last thing you need after your patented invention becomes a success is for someone to come out of the woodwork and claim that he or she is the real inventor or one of the inventors. Even if you bankrolled the entire operation, or came up with most of the ideas, you still have to identify minor contributors or subcontractors if they contributed to the features covered by your claims.

Don't insist on naming yourself as an inventor because of pride or ego, or improved job opportunities.

Inventorship is not the same as ownership. Just because someone is an inventor does not mean they have to share in the fruits of the patent. This can be taken care of with agreements you enter into with such people in which they assign all their rights to you or your company. But getting an assignment from an inventor or co-inventor does not mean that he or she is no longer an inventor. It just means that he or she is giving up all of his or her rights in the invention to you or your company. Your patent attorney

TIP

INVENTORSHIP AND OWNERSHIP

If you work in a corporate environment, or are a subcontractor or consultant, when you are hired you will probably be asked to sign an agreement stating that any inventions you create or contribute to will belong to your employer or the company hiring you. Read these agreements carefully. You should make sure that any inventions that have nothing to do with your job and which are created prior to employment and/or outside work can still be owned by you. And if you are a consultant working with a number of different clients, you should make sure that the scope of your task has been carefully defined in the contract. You don't want to end up in a situation where two of your clients are telling you that you are obliged to assign the same invention to each of them. You should definitely have a lawyer look at all these agreements so that you fully understand what you are giving up. And if you are pretty good at what you do, you might be able to negotiate an agreement where you get to have a piece of the action from any patents that result.

can walk you through this. You would think that this would be simple. But egos, jealousy, and competitive spirits can get in the way in this process. Make sure you do not let this happen to your invention, or everything you have worked for could be lost.

Filing for foreign protection

If you want to extend your patent to other countries, it is worthwhile knowing that there are various international agreements that can help you. These include the Paris Convention, the Patent Cooperation Treaty (PCT), and the European Patent Convention. The Paris Convention allows you one year to extend your patent claim to foreign countries with the same US filing date. And also, by filing a single international or PCT application, the decision to incur expense can be effectively put off for a further 18 months. Having this extra time after you file in the United States allows you to better assess the merits and potential value of your invention before incurring the

Inventorship is not the same as ownership. Just because someone is an inventor does not mean they have to share in the fruits of the patent.

significant costs of international patent protection. The European Patent is a particularly cost-efficient procedural tool for obtaining patent protection in different European countries.

If your invention is created in the United States, you need an express foreign filing license from the US Patent and Trademark Office before you can file any patent applications outside the United States unless you wait until six months after you file your US application. This is to give the government an opportunity to determine whether your application discloses any materials that could threaten the national security of the United States. This applies even if the subject matter of your patent application is for underwear. When you receive the filing receipt for your application, it will indicate whether the foreign filing license has been granted.

Every once in a while a US inventor wants to file the first patent application outside the US. You need to file a special request to get a foreign filing license to do this.

If you are up against a one-year deadline because of prior disclosures you made, you may need to get your application on file without a signature.

THE PATENT APPLICATION PROCESS

You have taken care of the basics. With the help of your patent attorney, you have gotten your full application on file. You paid attention to everything and now it's time for the US government to live up to its end of the deal. But remember, it's the US government. The serial number they give you for your application is like one of those numbers you get at the deli. Pack your lunch—and breakfast and dinner too—for 1–3 years. Even if your patent application is for a type of sandwich *(good luck!)*, you still need to wait in line.

The back and forth process with the United States Patent and Trademark Office is known as the "prosecution" of your application. Don't worry, you don't go to jail if your application is refused. It's a sort of stodgy term, so we won't use it again.

Preliminary questions

If you are up against a one-year deadline because of prior disclosures you made, you may need to get your application on file without a signature. Or it might end up that your patent application actually claims two completely different inventions. Inventors and their patent professionals sometimes realize this just as they are about to file an application. They want to get on file, so they put everything together in one application to get a filing date.

The Patent and Trademark Office will accept applications without signatures or which claim a couple of different inventions. But before they will fully examine it, they ask that the inventor or inventors sign it. And if there are a couple of different inventions, the Examiner will issue a paper or call your patent professional and tell him or her that you have to limit your application to just one of these inventions and move the other inventions to a different application. You will be given a fairly short period of time, a month or two, to straighten out these issues. For example, if there are two separate inventions, you will decide which one you want to go forward with. As to the other invention, you can either drop it or file another application, which will still have the same filing date, but requires you to pay new fees, or you can just wait to see what happens with the first invention you claim.

Publication of your application

For the first 18 months after you file your full application, it will be treated as confidential information and no one from the outside can access it. At the 18-month anniversary, your full application with claims will be published. You can ask that it be published earlier. Unlike trademarks, no one can object to this. It just lets everyone, including your competition, know what you are up to.

Office Actions

Once all the preliminary questions are out of the way, eventually you will enter into a dialogue with a Patent Examiner at the US Patent and

TIP

MAINTAINING CONFIDENTIALITY

If, after you file your application, you decide for some reason to abandon it, do this before the 18th month passes. The substance of your application will then remain confidential in the Patent and Trademark Office.

Trademark Office through what are called Office Actions. The Examiner will typically question different aspects of your patent application, and he or she might cite prior patents or other prior materials and conclude you aren't claiming anything new. Or, more likely, he or she will say that what you claim is obvious based on the prior materials.

This is where you will have an opportunity to tweak your claims or argue with the Examiner. If the claims in your application are drafted too broadly in order to cover a wide range of technology, you will probably encounter more obstacles from the Examiner than starting with narrower claims. It's tempting to limit your claims in order to have your patent application allowed, but often it is better to file written arguments to the Examiner explaining why what you claim is different from prior materials. This is the delicate balance that patent professionals are trained for. And the more they understand your true invention, the easier it will be for them to decide when to argue and how to scale back your claims. This back and forth can go on for a bit. Don't lose heart; stay cool and persevere—it happens to virtually all patent applicants.

If the claims in your application are drafted too broadly in order to cover a wide range of technology, you will probably encounter more obstacles from the Examiner than starting with narrower claims.

...often it is better to file written arguments to the Examiner explaining why what you claim is different from prior materials.

Patent Examiners at the US Patent and Trademark Office are, on the whole, quite accessible and, as with all good things in life, everything starts with communication. If you do not have a patent professional acting for you, do not hesitate to call up the Examiner or, if possible, meet with him or her. This helps you pinpoint the issues and facilitates resolving them. Work, as it were, with the Examiner rather than against him or her.

Notice of Allowability and other options

More often than not, your patent professional will be able to work out claims that are acceptable to the Examiner. You will also need to get other formalities out of the way. For example, the Examiner might ask that certain revisions be made to your drawings so they are consistent with your written description. Or you might need to change the drawings a little. You are allowed to do this as long as none of the changes you make introduce new materials or ideas to your application. If that happens, you may be forced to file a second application based on your first application. The new material you add will be given a new filing date of your second application. You should make certain that the new material you add has not lost the ability to be patented due to prior disclosures you made or the inventors might change. Speak to your patent professional openly and honestly about all this. Otherwise, you can end up with a vulnerable patent.

You should make certain that the new material you add has not lost the ability to be patented due to prior disclosures you made.

Sometimes the Examiner issues a "final" refusal if he or she does not agree with you. At this stage, you need to either appeal or extend your application by paying extra fees. This continues the examination process and gives you additional opportunities to submit written arguments or to change your claims further. If you decide to pursue an appeal, make sure you understand how much it will cost and what your chances of success are.

If all your homework was done and your invention is truly new and not obvious over what was already out there, the Examiner should finally allow your application. He or she will issue a paper called a "Notice of Allowability." Read this carefully, since sometimes there are a few final requirements the Examiner will want you to make, perhaps to tweak your drawing a little. You have three months to pay your fees to the Patent and Trademark Office. Once that is done, your patent will eventually be issued. Before that happens, however, remember to file any new applications for other inventions that were disclosed in this application. Or, if you want to add new matter and use this application as a jumping-off point, get this done now. Once your patent is issued, it can be cited against any new applications you file.

Issuance and maintenance

On the big day when you receive your Certificate of Patent Grant, pop a bottle of champagne. After all the hard work, you deserve it. But set aside some more money. You need to pay maintenance fees during the life of

Sometimes the Examiner issues a "final" refusal if he or she does not agree with you. At this stage, you need to either appeal or extend your application by paying extra fees. This continues the examination process and gives you additional opportunities to submit written arguments or to change your claims further.

your patent, or it will die a premature death. Otherwise, the natural life expectancy of your patent will be 20 years from the date you filed your full application with claims unless there were certain types of delays, in which case the patent term can be adjusted.

You need to pay maintenance fees during the life of your patent, or it will die a premature death.

WHEN DOES SOMEONE INFRINGE MY PATENT?

It is not at all uncommon for competitors to copy an invention before the patent has even issued. You can determine whether the claims as you have worded them cover what your competitor is doing. If they do not, there may still be some time to alter your claims to pick up what the copycat is doing.

However, until a patent is actually issued, you cannot sue anyone to stop them from using, making, or selling what you claim as your invention. There are procedures in the Patent and Trademark Office that you can use to expedite your application if you believe that your invention has been copied. Sometimes a demand letter can be sent out to your competitor before your patent is issued. You can warn your competitor that you are in the process of receiving a patent and that you are prepared to initiate litigation as soon as the patent issues. If the application has been published, you can send out

You can warn your competitor that you are in the process of receiving a patent and that you are prepared to initiate litigation as soon as the patent issues.

a copy of your claims so your competition knows they are in trouble. You might be able to work out a settlement so that, by the time your patent issues, the product is off the market or you have a license in place.

If you wait until your patent issues to send out letters or an infringement occurs after your patent issues, you still have to make sure your claims cover what your competition is doing. If you are willing to settle for a reasonable royalty, your opponent might be agreeable to something. But this is where we need to give you a reality check.

Patent litigation is incredibly expensive and can sometimes drag on forever. It also takes you and your employees away from your business. Instead, you end up spending a lot of time with your lawyers. Also, the first thing most opponents will attempt to do is to invalidate your patent by attacking its merits. They will say you did not disclose prior art. Needless to say, having a skilled lawyer at this stage is essential. Settling with your opponent is very often the sensible outcome in many a patent dispute. See chapter 7, pp232–255, for some more thoughts from us on litigation.

Patent litigation is incredibly expensive and can sometimes drag on forever.

Sometimes your competition not only infringes your rights but actually introduces a product with some improvements to it. This can put you in a bind.

When you send a letter to your opponent, he or she will generally respond with the types of arguments we have outlined above. But in some cases, an aggressive opponent will use your letter as a basis to start a lawsuit against you. This is called a "declaratory judgment" action. You might ask how that can happen; these are your rights that have been infringed. Well, it can. The other party can ask a court to declare that your patent is invalid or that he or she is not infringing your claims. For this reason, some patent owners choose to sue people before they even send a warning letter. Consult  with your lawyer carefully on this strategy. But remember, owning a patent does not mean that you cannot be sued.

Settling with your opponent is very often the sensible outcome in many a patent dispute.

Sometimes your competition not only infringes your rights but actually introduces a product with some improvements to it. This can put you in a bind. You know that consumers would rather have your opponent's product than yours. But your opponent needs a license from you to sell the product, even with the improvement, if the product incorporates the product you claim. If, in the meantime, your competitor has also received a patent on the improved product, you can be precluded from making that improvement to your own product. This is where a cross-licensing arrangement can come into play. Both sides take a license from the other. Cross-licensing opportunities come up even when there is no litigation. Think of the successful Sony and Samsung venture.

GENERATING REVENUES FROM MY PATENT

A single inventor in today's world will find that it can be very costly to manufacture or distribute a newly invented product alone. Often the best way to exploit your invention is to hook an 800-pound-gorilla company that shows an interest in the invention. You can either sell your invention and assign your patent to such a company for a lump sum or license your patent against a royalty income stream to a business partner.

Carefully weigh all the pros and cons of self-manufacture, selling, or licensing, but do not forget to include the factor of patent maintenance and potential defense costs into the equation.

You can also pledge or "mortgage" your patent as loan collateral to a bank or other lending institution.

CHAPTER 6
design patents

WHAT IS A DESIGN PATENT?

In the chapter on patents, we explained that generally, when everyone refers to "patents," they mean "**UTILITY PATENTS.**" Utility patents protect the way things work and are made. But what about the way things look? The shape of your **CELL PHONE** isn't entirely determined by the way it works. It also has a certain design, which may have caused you to buy it instead of another one. Look at your **COFFEEMAKER**. It probably looks different from

In today's marketplace, there are many competing products and everyone wants a piece of the action. New products are flooding into the marketplace from every corner of the globe. Technology is moving in many different

competitive **COFFEEMAKERS**. They all do the same thing, but they look different. Look at the **SHOES**, or **SLIPPERS**, or **SANDALS** you are wearing—they probably look different from whatever is on the feet of the next person you see after you read this sentence. Look at your **WATCH**—the shape of the **WATCH CASE**, the **FACE** of the **WATCH**, even the **STEM** of your **WATCH** might have **DESIGN FEATURES** that have nothing to do with the way the **WATCH** works. Instead, these are features that help distinguish the way one item looks from another.

directions at the same time. As a result, if you are introducing a new product, you need everything you can think of to distinguish your product from your competitor's. Certainly having a better product with new and

Even a simple screwdriver may have certain ornamental and decorative features added to it to give it a different look.

improved features is critical. Utility patents will protect that. A neat branding campaign will get your product attention. Trademarks will protect that. Good advertising copy and maybe even a theme song for your product get protected by copyrights. But all that may not be enough. You want your product to look good, too.

To get that extra edge, more and more people and companies devote their creative resources to the external design of their product or packaging—its appearance and eye appeal. This is one place where design patents come into play. Design patents are important to most fashion and luxury products. Product design is an integral part of the image of luxury and fashion houses.

If you are a product designer, an industrial designer, a package designer, or a company or individual getting ready to launch a new product, or if you are a designer of luxury or fashion products, you will have to understand what a design patent is, how you obtain one, how it differs from other types of intellectual property protection, and what you can do with your design patents.

Examples of designs protected by design patents

The Statue of Liberty, the soles of many **NIKE** shoes, the headlight on a **JAGUAR** car, a **GUERLAIN** perfume bottle, the icons in computer software, an ornamental necklace, distinctive packaging, the shape of a cookie, the shape of a new handbag.

WHAT DO DESIGN PATENTS PROTECT?

Design patents protect the ornamental and decorative aspects of any product or other article that can be manufactured. This includes the packaging that is used to sell a product. The protection extends to the shape of an article, three-dimensional designs applied to an article, and even two-dimensional ornamental designs applied to an article. The simplest designs can still have design patent protection. It does not need to be ornate—it just needs to be a feature that does not contribute to the way a product or package works, but instead to the way it looks. If the new features of your product or package serve a function, you should look to the utility patent laws to see whether that functional feature is new and not obvious over what was already out there. But if the feature has an original visual appearance, you are into the realm of design, which can be protected under the design patent laws. Even a simple screwdriver may have certain ornamental and decorative features added to it to give it a different look.

A design patent gives you a claim to exclude others from making, using, or selling your design as depicted in the drawings of your design patent. Design patents last for 14 years from the date they issue. There are no maintenance fees to pay along the way, and they cannot be renewed. Once your design patent expires, the design is there for anyone to make, use, or sell, unless it has become a trademark, which can happen in very rare instances. So get the most out of your design patent while it lasts.

...don't be surprised if one of your designs becomes a staple and everyone starts associating that product design with your company.

Make your design patents sweat

Yes. This is not a typo. We say make them sweat, not make them sweet! Design patents can work hard for you in an important way. In the chapter on trademarks, we explained that certain product and package designs are not automatically protectable. These designs generally need to be used in a certain way and promoted for a long time before consumers recognize them as brands associated with a single company. So, if you find yourself in that situation, how can you keep others from copying your product or package design while you are trying to build up trademark rights? Here is where bundling your rights and having a holistic approach to your intellectual property pay off. A design patent will protect the ornamental features of your product or package design as soon as the design patent is issued. Then, while your design patent is running and protecting the ornamental features of your design, through the use of good advertising campaigns you can teach your consumer base that they are not just looking at a pleasant design. They are looking at a product or package design that comes with your brand only. And don't forget, some features of your design might also be protectable under copyright laws.

This bundling of rights or leveraging from different types of intellectual property protection can be very useful if you have long-term plans for your designs. But if you are in a field where new styles are coming out every season, concerning yourself with building trademark rights in your product designs may not be relevant. But don't be surprised if one of your designs becomes a staple and everyone starts associating that product design with

SPEED THINGS UP

It can take a year or more for a design patent to be issued, but you can expedite it, which can cut down the time to maybe nine months. Once your design patent is issued, you have the ability to stop your competitors from copying based on your design rights. But what if someone copies your product beforehand? You are out of luck, although you can warn them (see p.228). This is one reason why you should to file your design applications as soon as your designs are completed.

your company. If that happens, and if you promote the design as a brand, you can take advantage of the 14 years your design patent gives you to build a reputation in your product design as a trademark. Then, by the time your design patent is ready to retire, you will have trademark protection. Remember, like diamonds, trademarks can last forever!

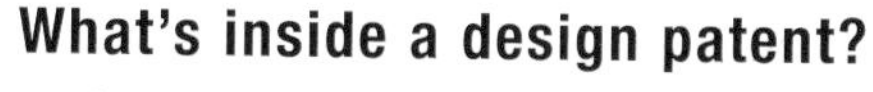

What's inside a design patent?

Unlike utility patents, design patents are very easy to read, or at least at first glance. Visit www.fromedisontoipod.com for an example. Information on the inventor and sometimes the companies that have assigned their rights appear on the cover, along with a drawing, which is usually a three-dimensional-type view of the object that is protected. As you flip through the pages, you might ask, where is the written description? Where are the claims? Well, in a design patent you are looking at them. There are no written claims in a design patent. The drawings are your claims. A picture is worth a thousand words and the drawings shown in your design patents should reflect all the design work you have done. The various drawings have to show your designed article from every angle—not just a three-dimensional view. This means that the drawings in your design patent application need to line up with one another—every element of your design needs to be consistently shown and clearly depicted. You can't leave anything to the imagination. After all, the US government is giving you exclusive rights in your design for 14 years, so everyone has to understand exactly what it is that you claim is your design.

Just as with utility patents, the key here is that you have only one year to test the market for interest in your design before filing a design patent application.

What are the basic requirements for a design patent?

Just like utility patents, you cannot get a design patent on any old design. Here is where design patents and utility patents are the same—to get a design patent, the features of your design need to be a) new and b) nonobvious over everything else that was out there already. But you cannot get a design patent for a feature of a product that is useful or functional. This is where design patents differ from utility patents.

New This means that your design has to be new or novel, compared to all the information and materials that were out there before your design was complete, and even that is not enough—you have to be on file within one year after the date of any publications you make of your invention anywhere in the world. And, just as with utility patents, this applies to your own activities. If you sell or offer for sale a product in the US that contains the design you want to include in your design patent application, you have to get on file within one year after you do that, or you are out of the running for a design patent. This also applies to any public uses of products with your design that you make in the US. Again, you have only one year to file.

And here is where there is another important distinction between design patents and utility patents. If you are based outside the United States and you have filed for design protection in another part of the world, and if that design protection is issued soon after you file, you need to get on file in the

THE DESIGNER STANDARD

Just as with utility patents, the standard is whether your claimed invention is obvious to another person skilled in your field. Remember that dude? Well, when getting design patents, this dude is a designer.

United States within six months of your foreign filing date or your ability to obtain a design patent in the United States will be lost forever.

Just as with utility patents, the key here is that you have only one year to test the market for interest in your design before filing a design patent application. So keep track of your activities! And remember, for designs, we encourage you to file your applications as soon as the design is complete. This will get the protection process rolling and help if someone ends up copying your design.

Nonobvious Just as with utility patents, your claimed design cannot be an obvious modification of what was already out there before your invention. Examiners can cite combinations of prior patents and pictures of other articles or even ads for your own earlier designs. They might tell you that it is obvious to come up with the design features you are claiming based on what is already out there and refuse to allow your application.

...you cannot get a design patent for a feature of a product that is useful or functional.

If you are trying to build up your design as a brand and you do not get your design patent application on file, you may not have any intellectual property protection.

WHAT SHOULD I DO TO PROTECT MY DESIGN?

Here, if you have not already read Chapter 5 (pp.164–205) you should read it now. The steps we recommend for design patent protection are basically the same as the steps we recommend for patent protection.

Get thee to the Patent and Trademark Office—and pronto!

First and foremost, get to a patent professional as soon as your design is complete. Don't make the mistake of thinking that you can file for design patent protection after you have had an opportunity to determine whether the design will be a success. As just explained, design patents follow the same rules as utility patents. Think of the one-year clock with the 10-ton weight hanging over your beautiful design (see p.175).

If you do not file for a design patent, you can have real exposure. If you are trying to build up your design as a brand and you do not get your design patent application on file, you may not have any intellectual property protection for a long time. Filing for design patent protection is a critical step that many companies forget to take or do not even know they can take.

Some designs come and go from the marketplace so quickly that it would be impossible to obtain a design patent while the article is still for sale. In our view, the best way to deal with this is to get your design patent application on file as soon as possible after the design is complete. Design patent applications are not nearly as expensive to file as utility patent applications. If you sit around and wait to decide whether the design will have any economic viability, you will have lost precious time if it does.

NOTE OF CAUTION!—Design theft

As soon as your designs are displayed at a trade show, don't be surprised if people take photographs of them or jot down notes, getting ready to copy them. The stealing of designs is so common that some people think there is nothing wrong with it. And there isn't if the creator of the design fails to seek appropriate protection.

Getting a design patent application on file before you make any display can enable you to tell your competitors and any people looking at the design that it is the subject of patent protection. You can use "Patent Pending" on the actual article, on trade show displays, on advertisements for the product, and on your website. This will let people know that you mean business and that they could be in trouble if they copy your design.

One of the beauties of design patent applications is that they are not published prior to issuance, unlike most utility patents. The first time your competition will see the features you claim are your design is when your design patent issues. If you have been smart and careful in the way you claim your design, it will be broad enough to pick up copycats.

TIP

PROVISIONAL APPLICATIONS

One more difference from utility patents: there are no provisional design patent applications. This is not surprising, since your design patent application is just the claim, the drawings you submit. A "provisional" utility patent application, on the other hand, does not have to have claims. Remember, that it just provides the written description of the invention. But because design patents do not have a written description, it would not make sense to have any "provisional" applications for designs. This is all the more reason why you should get your design patent application on file as soon as possible.

The stealing of designs is so common that some people think there is nothing wrong with it.

Keep good records

Retain good written records of your designs as they are developed. For prototypes, keep a copy in safe custody or a photograph as a record for establishing the date and origin of your design. Don't disclose your designs to anyone until you file your application. If you do, assume your design is going to be copied. This means that you need to make certain that you are on file with your design patent applications before you start showing your designs to others.

Searches

Obtaining a good design patent search before you file will help you look at previous designs to give you an idea of which features of your design may have already been protected. Your patent professional can get the search taken care of and will know how to advise you when he or she gets the results.

If you claim a design element that is not critical to your new design and a copycat leaves it off, chances are that they will not be found to have infringed your rights.

For starters, you can use the "**IMAGES**" category on the **GOOGLE** search engine to pull up designs in your field. If your design is the sole of a shoe, type in "shoe soles." Or you can use "jacket design" or "lampshade designs" and so on. This type of search will probably pull up a number of existing designs, which will help you determine whether your design is likely to be viewed as new and nonobvious compared to what is already out there.

FILING A DESIGN PATENT APPLICATION

Make sure your design patent drawings show your design

You might think that this will be a no-brainer. Unfortunately, even a number of companies that obtain design patents and even some patent lawyers think the same thing. They just hand over the designed article to a draftsman and tell him or her to make a set of drawings showing all the views of the article and then they file their application. Why not? Isn't this what you are trying to protect, the entire design of the article? In some cases, it is possible that the entire article has design features all over it that are new and nonobvious.

But in the world of design it is not uncommon to launch a new product that incorporates certain features that have already been out there and others that are new. You might be able to get a design patent on articles with new and old features combined. But why get a patent that covers old stuff? It only gives your competition an opportunity to make a few changes to the old stuff plus incorporate your new design. Then they will not be infringing

If the photograph shows your designed object sitting on a table, and you do not tell the Examiner that the table is not part of the claimed design, he or she will assume it is.

your design patent. This is because what you show is what you claim. If you claim a design element that is not critical to your new design and a copycat leaves it off, chances are they will not be found to have infringed your rights.

So, just like anything else in the world of intellectual property, you have to think about what it is about your design that is new. That is what you want to protect. Then communicate that to your lawyer so that he or she understands and then can go about protecting just that feature in the application. For example, if you are trying to protect the design on a new perfume bottle, but there is nothing about the cap that is new, why include the cap as part of your design?

How do you go about protecting just certain features?

If you were to submit drawings showing only bits and pieces of an article, the Examiner or a court would not understand what the article was. They would refuse your application on that basis. As a result, the Patent and Trademark Office has come up with very detailed rules that define exactly how you can depict your product and package designs when you are filing a design patent application. For example, you can use dotted lines to show portions of an article that are not part of your claimed design. This way, the Examiner will understand the overall article your design is used on but he or she will also understand the feature of it that you claim as your design. Design cuts across almost everything in the marketplace, so the ins and outs and nuances of what to show and how to show it for each category of article

We discourage using photographs at all, and forget using color photographs unless color is an actual feature of your design.

will vary. Your patent professional and the draftsman you use should be familiar with these rules.

The Patent and Trademark Office will also accept black and white photographs of all views of the design you would like to claim if you mount these photographs on paper. But, unless every feature of your product is new, this is not the way to go. It sounds easy, but you get what you pay for. If you use a photograph, you are claiming everything in the picture as your design. Even if you are cool with that, you still have to watch your step. If the photograph shows your designed object sitting on a table, and you do not tell the Examiner that the table is not part of the claimed design, he or she will assume it is. The same applies to reflections in the photograph, trademarks that might be displayed on the product, and all other elements depicted in the photograph.

The Patent and Trademark Office is very literal with the "drawings" you submit, whether they are prepared by a draftsman or whether they are photographs. You can submit color photographs and claim color as an element of the design, but this should be done only when a particular color is a key feature. Remember, if you show that color as an integral part, your competitor can probably avoid infringement by copying the shape of the design but depicting it in a different color. We discourage using photographs at all, and forget using color photographs unless color is an actual feature of your design.

Once these styles have been offered for sale or used or shown in ads for more than a year, they are prior art, even if you created them.

The Patent and Trademark Office is very literal with the "drawings" you submit, whether they are prepared by a draftsman or whether they are photographs.

One way to see how other people in your field are protecting their designs is to do some patentability searching for design patents of big players in your field. You can search under the "assignee" field at the US Patent and Trademark Office database (patft.uspto.gov/netahtml/search-bool.html).

You will see how certain features of designs are being protected on competitive products. But don't assume that your competition has figured out the best way to protect its designs. Some companies are very sophisticated in this field and others are not. Use these prior design patents to give you ideas as to how to go about claiming your own designs. Use them as another resource or tool when you are putting together your own strategy. Since the drawings are what you claim as your invention, you have to show every angle of your design.

This way, the Examiner will understand what it is when he or she is looking at the prior art to determine whether the claimed design is new and nonobvious. Likewise, your competitors can look at the issued design patents and understand what they can do and can't do. And a court will look at your claimed design to determine whether a third-party copycat has infringed it or not.

Other requirements to keep in mind

Since design patents are a type of patent, all the other rules we told you about in the utility patent section apply here too. So if you have not yet done so, you should read chapter 5 (pp.164–205) on utility patents as to invention disclosures, best mode, your continuing obligation to describe material prior art, and inventor and ownership issues.

With respect to your obligation and duty to disclose material prior art, for designs, we stress to you the importance of disclosing to the Patent and Trademark Office your own prior designs. This includes prior styles that you have sold or displayed, even if you have never obtained a design patent for them. Once these styles have been offered for sale or used or shown in ads for more than a year, they are prior art, even if you created them.

TIP

FOREIGN PROTECTION

If your design was created in the United States, just as with utility patent applications, you need a foreign filing license from the US Patent and Trademark Office before you can seek any type of design protection outside the United States. This applies even if a foreign country does not call the design protection "patent protection."

!

NOTE OF CAUTION!—One thing at a time

The Patent and Trademark Office will allow you to obtain only one design patent for a single creative concept. Sometimes, however, the single creative concept might be embodied in several related articles. For example, if your design has several unique elements that come together to create the design, or if some of the elements might be variable in the number of times they are repeated, you may be able to produce several sets of drawings depicting the same article incorporating these slightly different variations.

These are known as "alternative embodiments" and can be used to avoid filing several design applications on the same essential design concept. The differences between each embodiment, however, need to be very slight and the same theme and underlying design must be depicted in all embodiments. Otherwise, you need to file additional design patent applications for these designs.

Foreign protection

The United States is unique when it comes to design patent protection. Most countries do not consider ornamental designs the subject of patent protection. Instead, almost everywhere else designs are considered "industrial designs" and, in many instances, simple photographs of the designs can be filed. With all that said, a US design patent application can be extended overseas. But rather than having one year to file, you have only six months to file after your US design patent application is on file.

Once all of the preliminaries are out of the way, the most common questions that will arise will relate to your drawings, since these are essentially all that your design patent application comprises.

So make sure you understand whether you are getting foreign patent protection on a utility patent, where you have one year to file for foreign protection, or a design patent, where you have only six months.

THE DESIGN PATENT APPLICATION PROCESS

Once you file, you will receive a serial number. Design patent applications are usually examined sooner than utility patent applications. They are sometimes examined as quickly as several months after they are filed, but it can take up to a year and perhaps even a little longer for the examination process to begin. So don't be surprised if you do not hear from the Patent and Trademark Office for a long time after you get on file. When you do, here is what can happen.

Preliminary questions

Design patent applications undergo a preliminary examination and you will receive official actions from the Patent and Trademark Office if you have not submitted a signed declaration or if there are other basic questions relating to your application. It is also at this stage that the Examiner will look at your drawings to determine whether you may be claiming more than one basic design. This can come up if you have filed for alternative embodiments. The same steps can be taken here as with utility patent applications, where you can choose an embodiment and file divisional applications for your other designs. Or you can drop those other designs.

The Examiners are trying to make your patent application consistent with other design patents out there—this is generally a big help and not a hindrance.

Office Actions

Once all of the preliminaries are out of the way, the most common questions that will arise will relate to your drawings, since these are essentially all that your design patent application comprises. If you have used a good draftsman and have clearly shown what your design is, you might not even receive Office Actions and your application will be allowed. In fact, it is not at all uncommon for design patent applications to be allowed upon examination if good drawings are submitted. But if the examiner comes across prior art in searching or if the information disclosure statement you have filed shows certain prior art, the examiner might issue a refusal on the basis that your design is not new over something that was already out there or that your drawings are obvious over prior designs.

In our experience, encountering these types of rejections is rare. Examiners recognize that there are many designs out there and in fact some can be fairly similar. But they recognize that even the most subtle differences in designs can be a major advance for a particular product line.

Notice of Allowability and other options

Once everything is in order, you will receive a Notice of Allowability, so you can pay your issue fee and receive the design patent. But look at your Notice of Allowability carefully. Sometimes the Examiner has a small change to make. Some of these changes are made automatically by the

Examiner without asking you for permission. For example, he or she might change the title of your invention or they may change the manner in which you describe some of your drawings. Look over these changes carefully and, if you are not in disagreement with the Examiner, let them go. The Examiners are trying to make your patent application consistent with other design patents out there—this is generally a big help and not a hindrance.

Just as with utility patent applications, you have three months to pay your issue fees to the Patent and Trademark Office. Once that is done, your design patent will eventually issue. Make sure you file any divisional or other applications that you want to base this original application on before you pay your issue fee.

Examiners recognize that there are many designs out there and in fact some can be fairly similar.

When your design patent issues, you now have the exclusive right to make, use, and sell the features depicted in the drawings in your design patent. This right will last for 14 years from the date the design patent issues. One bonus is that, unlike "utility" patents, you do not need to pay any maintenance fees to keep it alive. Unlike logos or other design trademarks,

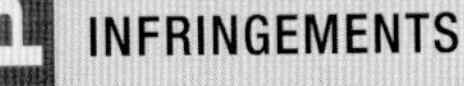

INFRINGEMENTS

Keep in mind that there is no obligation for you to sue every infringer. Unlike the situation with trademarks, you can pick and choose at random which infringers you wish to sue. Since litigation is expensive, you may prefer to go after the big groups and leave the small ones alone.

you do not have to use the design during this period. On the other hand, if you are trying to develop trademark rights in your design, you should be using it—and using it a lot.

WHEN DOES SOMEONE INFRINGE MY DESIGN PATENT?

LEGAL TEST—Infringement

The standard that the courts use in determining whether or not a design patent is infringed is whether the accused design is substantially the same as the claimed design. The focus is on deception of the ordinary observer, where one design is confused with the other.

As we have already warned, and as you may already know from bad experiences, people will copy your design the moment they see it. If that happens to you, and you have a design patent application on file, you should see your lawyer and consider sending such people a warning letter. You can also try to scale back what your drawings claim if you realize what you claim is more than what has been copied. You do not need to tell them when you expect your design patent to issue. If they know, they might try to dump as much product as possible in the marketplace before your design patent issues. If you do not let them know, and keep in mind that they cannot figure it out unless you tell them, since your design patent application is confidential and is not published, they will move forward at their own peril.

Don't focus on only one aspect of your intellectual property.

If they do not stop, remember, you cannot sue someone for design patent infringement until your design patent is issued. Once it does, you should sit down with your lawyer and be comfortable that your competitor's article incorporates your claimed design. Here is where having been very careful with what you show in your drawings and what you don't show as your claimed design will be essential.

If your competitor has added something else to the article in an effort to make it look different, that does not mean that he or she has not infringed your design patent. He or she still copied what you claimed. In fact, there may have been improvements to your design. That's fine, but it does not mean that your competitor is allowed to copy what you have protected. But if your claimed design has certain features that your competitor has not used, then you are probably out of luck, unless you can make a filing called a "reissue" to get your drawings changed to cover the copycat.

If your competitor has added something else to the article in an effort to make it look different, that does not mean that he or she has not infringed your design patent.

If you and your lawyer believe that the copied device infringes your design patent, a strong letter should be sent demanding that all sales of the design be discontinued. You can also ask for damages based on any manufacture, sale, or use of the articles made with your claimed design after your design patent is issued.

And if you have been selling your product at this stage for a while and it has been successful, you may also be able to claim that you have trademark rights in it. This is where approaching your intellectual property holistically is essential. Don't focus on only one aspect of your intellectual property. Think through all of the intellectual property that may have been infringed by your competitor. Does their design incorporate any of your trademarks, logos, trade dress, design, or even product design that has taken on trademark significance? Are there patterns or other features that might be protected under the copyright laws? Is there a way that the article works that you have protected or are in the process of protecting under utility patents? Is this a competitor who you think has found out critical information about trade secrets by hiring a former employee? It is all of these types of questions you should be thinking about before you send your letter. Then, send your letter.

If you have a reasonable opponent, he or she will realize the error of their ways. You might get a reply containing some obnoxious statements taking pot shots at your rights or even a lawsuit to declare there is no infringement or that your patent is invalid. But if the opponent says anything in the letter that indicates a willingness to settle, then in our view it is better to try to settle rather than litigate.

Just as with your utility patents, you are allowed to license and mortgage your design patent to generate revenues.

If you have a reasonable opponent, he or she will realize the error of their ways. You might get a reply containing some obnoxious statements taking pot shots at your rights. But if the he or she says anything in the letter that indicates a willingness to settle, then it is better to try to settle rather than litigate.

CHAPTER 7

intellectual property portfolio

WHAT IS AN INTELLECTUAL PROPERTY PORTFOLIO?

If you have read our chapters on the categories of **INTELLECTUAL PROPERTY**, you will appreciate that it is not at all uncommon for **A NEW CREATION** to have different aspects of it protected by different types of **INTELLECTUAL PROPERTY**. We have encouraged you to think in terms of bundling your **INTELLECTUAL PROPERTY RIGHTS**. Look at any new **CREATION** from an

Step back and look at your options for protection. As a whole, they give you an opportunity to build a fortress around your creation so you can maximize the benefits you can get from it. You are also making business decisions when you choose not to pursue intellectual property protection: you might miss opportunities if you ignore taking certain steps from the beginning. Remember the 10-ton weight with the one-year clock ticking over your invention? And what if someone else files a trademark application for a similar mark you plan to use but haven't gotten around to filing for?

ORGANIC or **HOLISTIC PERSPECTIVE**. It's not a good idea to get hung up on one aspect of **PROTECTION**. In fact, it's pretty rare that something new can be protected in only one way. That doesn't mean that you need to spend thousands of dollars pursuing **ENDLESS PROTECTION**. But it does mean that you should be aware of your **OPTIONS** and then make wise **BUSINESS DECISIONS** about what to **PROTECT**.

That could cause you to change your entire branding scheme or pay that person a lot of money to get them out of the way. By taking steps to protect your new creations, you better the chances of an infringer being caught or staying away in the first place. Believe us, if you get into any intellectual property battles, you will need strong and varied protection. This is true whether you are involved in the full-scale war of an infringement suit or the more subtle art of settlement negotiations.

...if you get into any intellectual property battles, you will need strong and varied protection.

Now, we hope you can look back at the **PANERAI** branded watch example on pp.40–41 and appreciate how a single item can be protected by different categories and layers of intellectual property. That example showed some of the more obvious examples. They are repeated here, but the list has grown to show how the possibilities are endless. At least 17 forms of intellectual property could be available for such a watch:

- Trademark rights in the house mark **PANERAI**.
- Trademark rights in the name **LUMINOR MARINA**.
- Trademark rights in the three-dimensional product design for the watch.
- Trademark rights in the three-dimensional product design for the crown lock.
- Design patent rights in the watch-face design.
- Design patent rights in the watch-case design.
- Design patent rights in the lock protector.
- Design patent rights in the bezel.
- Design patent rights in the watch package.
- Trade secret rights in the manner of assembling the watch mechanism.
- Trade secret rights in the manner of manufacturing some of the parts.
- Utility patent rights in the new and improved features of the watch mechanism that cannot be reversed engineered or protected by utility patents.
- Utility patent rights in the way the watch box is made.
- Copyright in the design drawings of the watch.
- Copyright in advertisements that feature the watch.

- Rights of publicity if any personalities are used to endorse or promote the watch.
- Trade name rights in the company name **OFFICINE PANERAI.**

As you can see, a seemingly simple product like a watch, its packaging, and advertising campaign can be protected by intellectual property in numerous ways. Trademarks, design patents, and utility patents are all spilling over. Trade secrets and know-how play an important role and copyright is there to protect the artistic expression on the underlying design drawings and any advertisments. At the manufacturing level, trade secrets and utility patents protect the manufacturing equipment and packaging processes and there are copyright and other patent rights for the software that tracks inventory.

If this list of possibilities for bundling your intellectual property starts to look endless to you, it is. The companies and individuals involved in the creation, design, manufacture, and distribution of consumer products have substantial investments in place.

Whether you are buying a tube of **COLGATE** toothpaste off the shelf at a **CVS** store or downloading the newest **GREEN DAY** release for your **iPOD** player, you can be assured that there have been a lot of people involved in getting these products and services delivered to you so that you know you are getting the best product for your money. And a long list of intellectual property rights is available to protect the labors of these people and companies.

Is there a chance that the product design of your new lawn mower could become a trademark someday? If so, start to promote the product design itself as a trademark.

HOW CAN YOU RECOGNIZE WHAT YOU HAVE?

Let's say you have just come up with a great new functional feature on a lawn mower. You know you want to file a utility patent application. But here are some other questions to ask before launching your lawn mower or shopping around:

- Is there anything about the design of the lawn mower that warrants design patent protection?
- Have you done searches of utility patents and design patents to see what else is out there? This could impact the way you state your written claims or submit your drawings.
- Have you done searches of existing patents to see whether your lawn mower will infringe anyone else's rights? You might have a patentable improvement on one feature but the underlying feature might be covered by another patent. If so, should you maybe approach the owner of the patent to offer a cross-license? Make sure to have them sign a nondisclosure agreement first.
- What is the brand you will use to sell this? If this is a new brand, have you searched it? Or is this a line extension where you already have coverage for certain types of garden equipment like hedge trimmers but not a lawn mower? If that is the case, you may need to file a new trademark application. And you had better still do a trademark search to see whether anyone else has rights for this new product. In doing so, you might discover a newcomer to whom you should send a cease-and-desist letter to protect your existing rights and to make sure the competitor doesn't think that he or she can stop your expansion. And be ready to negotiate.

- And what about a logo? Do you have one? Do you want to register it as a trademark? Is your logo a pretty design that might be the subject of a copyright registration? You should also make sure you own any copyrights in your logo. If it was created by an outside agency or individual, you need an assignment. The same applies to all of the advertising copy that's created.
- Speaking of ads, is there a chance that the product design of your new lawn mower could become a trademark someday? If so, start to promote the product design itself as a trademark. It takes a long time to turn most product designs into trademarks.
- Are you planning to use a still from the movie **EDWARD SCISSORHANDS** in your ads to market your lawn mower? If so, you need a license to reproduce the still from the copyright holder. And if **JOHNNY DEPP** is shown, don't forget about his right of publicity. Just getting clearance to use the still doesn't give you the right to use his image for your commercial gain. And also remember that any alterations you make to the still will need another license from the original copyright holder. Having a copyright license to copy doesn't give you the right to alter the work—that's another right.
- Is there anything about the way you assemble the lawn mower that will give you a competitive edge? If that is part of the claimed invention, you will need to disclose it in your patent application. But if it is separate, do you want to file another patent application on this development? Or is it a technique you can keep secret as it can't be figured out by reverse engineering? If so, you might be able to protect it as a trade secret. Do you have appropriate agreements in place with employees who have access to this information, confirming that they will keep it confidential? Do you make visitors to your facility sign in and do you keep track of them while they are there, making sure that they do not visit areas where your trade secrets may be at work?

There is usually more intellectual property than meets the eye.

Think about what to protect

If you are about to throw this book against the wall after all these questions, calm down. The above questions give you options. They empower you to decide how best to protect your creation under the circumstances. Which ones to choose will vary depending on the planned life of your creation and what you think you can get out of it. There is usually more intellectual property than meets the eye. Following the guidance in the chapters on trademarks, rights of publicity, copyright, trade secrets, utility patents, and design patents, for each new creation it would be wise to go through a quick checklist (see the Checklist, p.46) to see whether your new creation qualifies for protection in more than one category of intellectual property. You may have a new creation that deserves protection in several categories of intellectual property and even multiple protection within an intellectual property category, such as protecting a trademark, a logo, a slogan, and perhaps even the color of your product.

As demonstrated by the questions above, in the lawn mower example, you also need to be thinking about other people's rights when you come up with new creations. Could you be infringing someone else's copyright? If so, what types of clearances do you need? What about design patents, utility patents? Where have your employees or consultants come from—could they be using trade secrets they learned somewhere that they are not permitted to disclose to you?

TIP

WHAT TO PROTECT

The cost and expense, financially and time wise, when you are assembling an intellectual property portfolio are always important factors, whether you are on your own or working with a partner or company. Do your homework well and approach the issue of intellectual property protection in the same way as you would any other important decision you are making. Ask your intellectual property professional for an upfront estimate before you start. Even filing fees, especially patent fees, can be very expensive. So ask for a full estimate—not only for the short term—but also for the long term. Maintenance fees in themselves may be expensive. Do as you would do with any other item in business: set up a proper budget and manage it during the different stages along with your professional.

!

NOTE OF CAUTION!—Insurance

While you want to build a good portfolio of assets that you can sell, license out, or even use to help get financing, you also want to be in a good position to defend any claims that third parties might make against your activities. Should you have special insurance (known as errors and omissions insurance) for your project? Is there any other type of insurance that you should be looking at, such as product liability insurance?

USING INTELLECTUAL PROPERTY PROFESSIONALS

Intellectual property is one of the most dynamic and constantly changing areas of the law. Like many areas of law, it is rife with potholes and loopholes. We have given you only an overview here. It is the tip of the iceberg. Each category of intellectual property is filled with special rules, exceptions, and nuances that only a professional will recognize and know how to sort out. As a result, many intellectual property lawyers focus their practice in just a few areas. However, most have a good handle on the overall field and can guide you to other attorneys when they cannot personally advise you on a particular matter or issue. And they can help you make good business decisions when it is time to decide which intellectual property rights to pursue.

Intellectual property is one of the most dynamic and constantly changing areas of the law. Like many areas of law, it is rife with potholes and loopholes.

Patent applications, whether they are for utility patents or design patents, can be filed by the inventors of that creation. That is a bad idea. It is better to use a "patent attorney" or "patent agent." Such people have qualified by taking a special exam in the United States Patent and Trademark Office. Patent agents are allowed to file and handle patent applications but they are not allowed to provide legal advice. A patent attorney is admitted to practice law, so he or she can advise you on any legal issues as well as filing and handling your application.

Any attorney is allowed to advise you on intellectual property matters, but he or she is not allowed to file patent applications. Thus, if you need to litigate your patent, sell your patent, require general advice in connection with your patent, or want to evaluate your overall rights, any lawyer can advise you. But make sure he or she understands the field and has experience with patents. In fact, there are many litigators who handle patent matters who are not patent lawyers.

For other categories of intellectual property, you do not need a patent attorney to make your filings. Trademark applications and copyright applications are frequently filed and handled by attorneys who are not patent lawyers, although we urge you to hire an attorney with experience in the field. In many corporations, highly skilled paralegals handle the trademark and copyright filings. This can work very well provided they are managed by an attorney.

Each category of intellectual property is filled with special rules, exceptions, and nuances that only a professional will recognize and know how to sort out.

Even intellectual property professionals develop expertise among themselves. There are specialists who will deal only with patents and others who will deal only with trademarks. If you are dealing with more than one specialist within a firm, it is very important that you ensure that they properly coordinate their efforts. This sounds easy but, believe us, managing and coordinating the efforts of specialists isn't always a walk in the park. But it usually pays off to work with a legal team.

It can be difficult to find an intellectual property attorney who fills all your needs when you are starting out with your business, especially if you are not certain which category or categories of intellectual property protection are available to your creations. While there are individual practitioners who are excellent and can provide you with a very personalized service, a law firm will have the expertise of a number of different attorneys available. But even within the world of law firms there is great variation. Some firms are known as "boutiques" because they focus their entire practice on the protection of intellectual property. And other firms are large firms with offices all over the country, or even the globe, with intellectual property departments.

Don't underestimate how important it is to have a good professional relationship with your attorney and his or her firm.

So who should you use?

At the end of the day, we believe it comes down to the chemistry between you, or your company, as the client, and the lawyer and law firm you choose. Don't underestimate how important it is to have a good professional relationship with your attorney and his or her firm. You need a good comfort zone with your attorney. Make sure he or she listens to you and understands what you want to achieve. Give your attorney the necessary factual details so he or she can understand what you have. This should include any problems that you know about.

You should view yourself and your business as a "consumer" of legal services. While attorneys do not like to be considered simple "vendors" and pride themselves on being professionals, there are plenty of intellectual property attorneys out there and you do not have to settle for someone you don't like or you don't think has the time for all your work. Don't be embarrassed to ask questions when interviewing a potential lawyer or law firm about hourly rates and other types of financial arrangements like fixed fees for certain projects. You generally get what you pay for and this is a field where you do not want to be a bargain-hunter. It can be very frustrating to work with attorneys who take forever to turn around paperwork and to get things on file. For this reason you will also want to lay down the law with your attorney about what you need, when you need it, and what your budget is. But be realistic—remember, your attorney is handling many matters for many clients and everyone cannot have everything all at once.

> **TIP**
>
> **CHECK QUALIFICATIONS**
>
> As always, a note of caution. There are firms out there who are less then reputable. These firms will offer you a smorgasbord of all kinds of services without being properly equipped to render them. You should check and double-check the qualifications and expertise of the intellectual property professional you select.

We live in a world of regulations. Many industries are subject to federal and sometimes local laws on how to label certain products. There are even special registries for certain industries and some industries require you to file a federal trademark application. The states have requirements about setting up corporations and other entities. Further, if you have a partnership, it's important to have a written agreement. You should have a good corporate attorney who can guide you through these areas. You may set up special entities such as limited liability companies or corporations or partnerships to own your intellectual property portfolio and other assets of your business. In doing so, you may bring in partners, investors, or other entities who will have a financial interest or some other type of interest in your intellectual property.

The moment other people or entities start taking on one of these interests, you need to start thinking about whether your own personal interests may be different from the interests of the ownership entity and you should let your lawyer know if this is beginning to happen. This can lead to what is known as a conflict of interest. Attorneys have ethical obligations to make certain that they avoid conflicts of interest when representing their clients. In some cases, attorneys can be forced to withdraw from representing all entities on a particular matter if a dispute arises. This could mean that an attorney who you have worked with for a long time and who knows your business can no longer represent you.

USING OTHER PROFESSIONALS

Intellectual property has become one of the most valuable business assets in today's world. Uncle Sam is very much aware of this fact as well. So be careful. If your intellectual property venture is taking off, there are some points to take into consideration. When selling your intellectual property creation or buying someone else's, or when licensing your own creation, careful tax consideration and planning will be required. If your business starts taking off in the international arena or you plan from the outset to go global, it is advisable to receive good advice on tax-efficient intellectual property ownership structures.

And if you have substantial assets or if your intellectual property is or has potential of being a substantial asset, you should find a good attorney who handles estates. These professionals can best advise you on the pros and cons of different types of ownership structures and how they may impact you, your business, and your heirs in the long run. Some intellectual property attorneys are adept at advising on such matters. Or they can at least help you identify some of the issues. But it is a good idea to have other professionals know about your intellectual property if it is at all a significant part of your life.

In addition to the above advisors, there are a variety of intellectual property service firms that do searches for trademarks, copyrights, or patents. Some of these firms are not law firms and cannot provide you with advice. They can provide you only with factual information. It is generally a good idea to rely on your attorney to decide which search firm to work with when handling patent searches or trademark searches. This does not mean that you should not try to compile as much information as possible on your own. The Internet has now given everyone access to phenomenal amounts of information, and you may very well know how to navigate through the Web to find information that even your attorney cannot readily ascertain. In fact, when you hire your attorney you should let him or her know about these sources. But this is not a substitute for using a good professional search firm.

Intellectual property has become one of the most valuable business assets in today's world. Uncle Sam is very much aware of this fact as well. So be careful. If your intellectual property venture is taking off, there are some points to take into consideration... it is advisable to receive good advice on tax-efficient intellectual property ownership structures.

TIP

EXAMINERS

Examiners and Examining Attorneys can sometimes issue refusals or raise questions that are inappropriate or simply wrong. When this happens, if it is pointed out with a good explanation, the Examiner or Examining Attorney will almost always drop the question or refusal. But, if it comes to a point where you think you are right, you can always ask to speak with their Managing Examiner or Managing Attorney. Don't be surprised if they side with the Examiner or Examining Attorney. But if you explain the situation and shed some light on the situation they might better understand the facts and then make a suggestion to help you get over whatever bridge you are stuck on.

RESPECT AND PAY ATTENTION TO PATENT EXAMINERS, TRADEMARK EXAMINING ATTORNEYS, AND COPYRIGHT EXAMINERS

As mentioned in the previous chapters, the Examiners in patent and copyright applications and Examining Attorneys in trademark applications will communicate with you or your attorney through what are called Official Actions. The key here is to establish a successful line of communication with your Examiner. Do not approach the Examiner as the enemy. We suggest that you call your Examiner and have a talk. You will almost always find the person you are dealing with to be helpful and professional. Find out what the real concern is and address it head on in a constructive way.

Keep in mind that it is the Examiner's job to analyze and examine your application. He or she will, therefore, raise questions in relation to the merits or technicalities of your patent, or trademark, or copyright application. Do not take these questions personally. Polite, concise, and common-sense responses are always the best way forward. The examination process is a necessary hurdle to be tackled in a professional and polite manner.

Do not approach the Examiner as the enemy. You will almost always find the person you are dealing with to be helpful and professional. Find out what the real concern is and address it head on in a constructive way. Keep in mind that it is the Examiner's job to analyze and examine your application. He or she will, therefore, raise questions in relation to the merits or technicalities of your patent, or trademark, or copyright application. Do not take these questions personally. Polite, concise, and common-sense responses are always the best way forward. The examination process is a necessary hurdle to be tackled in a professional and polite manner.

If you know from the outset that you will use your new intellectual property creation not only in the US but also in other countries, you should consider an international filing program.

FILING OUTSIDE THE US

If you know from the outset that you will use your new intellectual property creation not only in the US but also in other countries, you should consider an international filing program. Also, you may initially start off in the US and then, if your creation turns out to be a huge hit, decide to expand your business to other countries. In these circumstances, you should be aware that there are a number of international intellectual property treaties that allow you to use your US application as a springboard to extend to other countries.

In fact, quite a lot of work has been done by the World Intellectual Property Organization and other bodies in the last few years, which has helped establish regional and international filing systems.

NOTE OF CAUTION!—Some points to consider

A few points to take into account when contemplating an international filing program: certain intellectual property treaties allow you to maintain your crucial priority date in the United States when you extend your applications or registrations to other countries. This must be done within a six-month period (in the case of trademarks and design patents) and one year, which may in some instances be extended to two and a half years (in the case of patents). Typically, when using the international filing systems, your costs are reduced and you may even have the added bonus of having an expedited application procedure.

Let's say you would like to extend the sales of your product or service to Germany and the UK. Rather than filing trademark applications in these two countries separately, you can simply file an application covering nearly all of Europe at the Community Trademark Office in Alicante, Spain. If you wish to extend your patent rights to the rest of Europe, Japan, China, and a few other commercially important countries, you can make use of the international Patent Cooperation Treaty to accomplish this most effectively. These international systems usefully serve as a "one-stop shop."

Needless to say, the pros and cons of an international filing system and how to get the best bang for your buck, making use of them on the most efficient basis, should be worked through with your intellectual property professional.

After years of international haggling, a number of countries have now signed up as members of these international intellectual property treaties, which in practice means that you can extend your application or registration to quite a significant number of countries in the world. The relevant treaties are the Paris Convention, which has over 160 country members, the Madrid Protocol and Agreement, with over 75 country members, the Hague Convention, with over 60 members, and the Patent Cooperation Treaty, with over 130 country members.

WARNINGS AND MARKINGS

Once your filings are made and you can launch or publish your creations, you will want to put the world on notice of your intellectual property rights. In fact, if you fail to use the proper notices for issued patents and registered trademarks, your ability to get money damages in some cases can be impacted. So it pays to use good notices. As a practical matter, notices can also have a psychological impact on people thinking of ripping off your creation. It's a polite way of warning people not to mess with you. In the prior chapters we have provided more information, but here is a summary:

Copyright	► ©2003, 2007. Mary Doe. All Rights Reserved. (The first year is when the first version of this work was published and the second year is when the current version is published).
Design Patents	► Patent Pending (while your design patent is pending) ► U.S. Patent No. D123. (where 123 is your design patent number)
Patents	► Patent Pending (while your application is pending). ► U.S. Patent No. 123 (where 123 is your patent number)
Trademarks	► XYZ™ (before your mark is registered) ► XYZ® (after your mark is registered)

> **TIP**
>
> **KEEP UP TO DATE**
>
> Remember to update your intellectual property notices and other markings as the status of your intellectual property changes. The same applies to patents. Your patent attorney should provide to you the patent number of your patent just before it issues. You should immediately begin changing your labels and other markings to show the patent number. It is very frustrating for us when we have spent a lot of time registering a trademark for a client and then years later continue to see a ™ rather than an ® on product packaging. It also can affect your ability to collect monetary damages in lawsuits, so updating your legends can make economic sense as well as keeping your attorney happy.

Use these notices whenever your mark is prominently displayed on packaging and in advertisements and promotional materials. You should use it with your logos too. The notices do not need to go after each use on your materials. That looks geeky and is usually overkill.

If there are a variety of intellectual property rights contained in a particular creation, which is often the case, you can use legends that are prominently displayed usually at the back of packaging or on the product itself or on labels. The examples will vary depending upon what type of intellectual property is involved, but here are three, just to give you some ideas:

a **XYZ and the XYZ Logo are trademarks. All unauthorized use is prohibited. ©2006 John Smith. All Rights Reserved.**

b **Patent Pending. XYZ® is a registered trademark and the XYZ logo is a trademark. All unauthorized use is prohibited.**

c **Patent Nos. _______, _________, and _______, Design Patent Nos. ________, ________, and ________. ©2002, 2004, 2007. Mary Doe.**

If your product is being shipped outside the US, your lawyer can guide you on the other types of notices that are required in the countries where you do business.

TIP

USE A MEDIATOR WHEN NECESSARY

In our experience, mediation is a very useful method of resolving conflicts quickly and efficiently. A mediator is a neutral referee (typically an intellectual property professional) who serves as an objective go-between for both sides to solve the conflict professionally in a swift and inexpensive way. In fact, many courts are beginning to require that disputing parties try to resolve an intellectual property conflict with a mediator. Arbitration, which is a more formalized version, might also achieve the same result.

INTELLECTUAL CONFLICTS AND LITIGATION

Registries around the world are becoming ever more cluttered as people become more aware and assertive in their pursuit of their intellectual property rights. Many more conflicts and a rise in litigation have been the result. At some point you may very well find yourself on either the sending or the receiving end of an intellectual property cease-and-desist letter. This is part and parcel of intellectual property enforcement. It can be truly nerve-racking at times, but more often it is simply irritating. Make peace with the knowledge that you will not always have a smooth ride. However,  your intellectual property lawyer will navigate you through these treacherous waters and, if common sense and good business judgment prevail, you will probably survive. Allow us to quote Fred's father, who, after 40 years of law practice, gave him the following words of wisdom:

"When your client is looking for blood, settlement seems tame. But always, always try to settle first. Only if you've tried everything reasonable to settle out of court and failed, do you take off the gloves. And then: to battle."

So turn the other cheek before you think eye for an eye. Settling out of court will save you enormous heartache. It will save on outrageous expenses in legal fees but also, crucially, on valuable management time. Most cases are brought for commercial and legal reasons, but at some time during the proceedings the whole process will become unavoidably emotional and

...turn the other cheek before you think eye for an eye.

sometimes personal. Parties become entrenched in their positions and emotions run high. This is what we consider the point of no return. Try to practice restraint. Common sense can, and should, prevail in these situations: a decision to settle can be a viable and more sensible option. Of course, the other side may show no interest in being reasonable and common sense or good business judgment may not be a concept within their frame of reference. In these cases, the second part of Fred's father's advice applies: "Take off the gloves and to battle." Arm yourself with the best possible team. Pick your intellectual property litigators carefully, and good luck!

ONE FINAL THOUGHT (HUMOR US, WE'RE LAWYERS. WE CAN'T HELP IT)

ARTICLE I, Section 8, Clause 8 of the US Constitution states that the Congress shall have the Power: To promote the Progress of Science and useful Arts, by securing for limited Times to Authors and Inventors the exclusive Right to their respective Writings and Discoveries.

Along the way, the patent and copyright laws have developed to protect your inventions, designs and works of authorship. Other laws have developed to protect your trademarks, trade secrets, and rights of publicity. The laws are on your side and so is the support of our society and culture. Let your creative spirit run wild, and make sure to protect your creations along the way!

appendices

Please note that these Appendices are here to help and guide you only. They will not cover every possible legal situation. Some of the suggestions will not apply in all situations and modifications will be required. These examples do not take away from the need to get expert legal help. Neither the publishers nor the authors accept any responsibility for any losses that may arise from the use or adaptation of these sample agreements.

These documents are just examples. You should work closely with your intellectual property lawyer on any document before it is signed, in order to ensure that your rights are protected.

CLASSIFICATION OF GOODS AND SERVICES

The following headings give general information about the types of goods and services that belong to each class. The list is not complete and is a guide to the classes that may be required. The US Patent and Trademark Office has an excellent search tool to help you select specific identifications at http://tess2.uspto.gov/netahtlml/tidm.html. Goods are in classes 1 to 34. Services are in classes 35 to 45.

Goods

Class 1 Chemicals used in industry, science, and photography, as well as in agriculture, horticulture, and forestry; unprocessed artificial resins, unprocessed plastics; manures; fire extinguishing compositions; tempering and soldering preparations; chemical substances for preserving foodstuffs; tanning substances; adhesives used in industry.

Class 2 Paints, varnishes, lacquers; preservatives against rust and against deterioration of wood; colorants; mordants; raw natural resins; metals in foil and powder form for painters, decorators, printers, and artists.

Class 3 Bleaching preparations and other substances for laundry use; cleaning, polishing, scouring, and abrasive preparations; soaps; perfumery, essential oils, cosmetics, hair lotions; dentifrices.

Class 4 Industrial oils and greases; lubricants; dust absorbing, wetting and binding compositions; fuels (including motor spirit) and illuminants; candles and wicks for lighting.

Class 5 Pharmaceutical and veterinary preparations; sanitary preparations for medical purposes; dietetic substances adapted for medical use, food for babies; plasters, materials for dressings; material for stopping teeth, dental wax; disinfectants; preparations for destroying vermin; fungicides, herbicides.

Class 6 Common metals and their alloys; metal building materials; transportable buildings of metal; materials of metal for railroad tracks; nonelectric cables and wires of common metal; ironmongery, small items of metal hardware; pipes and tubes of metal; safes; goods of common metal not included in other classes; ores.

Class 7 Machines and machine tools; motors and engines (except for land vehicles); machine coupling and transmission components (except for land vehicles); agricultural implements other than hand-operated; incubators for eggs.

Class 8 Hand tools and implements (hand-operated); cutlery; side arms; razors.

Class 9 Scientific, nautical, surveying, photographic, cinematographic, optical, weighing, measuring, signaling, checking (supervision), life-saving and teaching apparatus and instruments; apparatus and instruments for conducting, switching, transforming, accumulating, regulating, or controlling electricity; apparatus for recording, transmission or reproduction of sound or images; magnetic data carriers; recording discs; automatic vending machines and mechanisms for coin-operated apparatus; cash registers; calculating machines, data processing equipment and computers; fire-extinguishing apparatus.

Class 10 Surgical, medical, dental, and veterinary apparatus and instruments, artificial limbs, eyes, and teeth; orthopedic articles; suture materials.

Class 11 Apparatus for lighting, heating, steam generating, cooking, refrigerating, drying, ventilating, water supply, and sanitary purposes.

Class 12 Vehicles; apparatus for locomotion by land, air, or water.

Class 13 Firearms; ammunition and projectiles, explosives; fireworks.

Class 14 Precious metals and their alloys and goods in precious metals or coated therewith not included in other classes; jewelry, precious stones; horological and chronometric instruments.

Class 15 Musical instruments.

Class 16 Paper, cardboard, and goods made from these materials, not included in other classes; printed matter; book binding material; photographs; stationery; adhesives for stationery or household purposes; artists' materials; paint brushes; typewriters and office requisites (except furniture); instructional and teaching material (except apparatus); plastic materials for packaging (not included in other classes); printers' type; printing blocks.

Class 17 Rubber, gutta-percha, gum, asbestos, mica, and goods made from these materials and not included in other classes; plastics in extruded form for use in manufacture; packing, stopping, and insulating materials; flexible pipes not of metal.

Class 18 Leather and imitations of leather, and goods made of these materials and not included in other classes; animal skins, hides; trunks and traveling bags; umbrellas, parasols, and walking sticks; whips, harness and saddlery.

Class 19 Building materials (nonmetallic); nonmetallic rigid pipes for building; asphalt, pitch, and bitumen; nonmetallic transportable buildings; monuments, not of metal.

Class 20 Furniture, mirrors, picture frames; goods (not included in other classes) of wood, cork, reed, cane, wicker, horn, bone, ivory, whalebone, shell, amber, mother-of-pearl, meerschaum, and substitutes for all these materials, or of plastics.

Class 21 Household or kitchen utensils and containers (not of precious metal or coated therewith); combs and sponges; brushes (except paint brushes); brush-making materials; articles for cleaning purposes; steel wool; unworked or semi-worked glass (except glass used in building); glassware, porcelain, and earthenware not included in other classes.

Class 22 Ropes, string, nets, tents, awnings, tarpaulins, sails, sacks, and bags (not included in other classes); padding and stuffing materials (except of rubber or plastics); raw fibrous textile materials.

Class 23 Yarns and threads for textile use.

Class 24 Textiles and textile goods not included in other classes; bed and table covers.

Class 25 Clothing, footwear, headgear.

Class 26 Lace and embroidery, ribbons and braid; buttons, hooks and eyes, pins and needles; artificial flowers.

Class 27 Carpets, rugs, mats and matting, linoleum and other materials for covering existing floors; wall hangings (nontextile).

Class 28 Games and playthings; gymnastic and sporting articles not included in other classes; decorations for Christmas trees.

Class 29 Meat, fish, poultry, and game; meat extracts; preserved, dried, and cooked fruits and vegetables; jellies, jams, fruit sauces; eggs, milk, and milk products; edible oils and fats.

Class 30 Coffee, tea, cocoa, sugar, rice, tapioca, sago, artificial coffee; flour and preparations made from cereals, bread, pastry, and confectionery, ices; honey, treacle; yeast, baking powder; salt, mustard; vinegar, sauces (condiments); spices; ice.

Class 31 Agricultural, horticultural, and forestry products and grains not included in other classes; live animals; fresh fruits and vegetables, seeds, natural plants, and flowers; foodstuffs for animals; malt.

Class 32 Beers; mineral and aerated waters and other nonalcoholic drinks; fruit drinks and fruit juices; syrups and other preparations for making beverages.

Class 33 Alcoholic beverages (except beers).

Class 34 Tobacco; smokers' articles; matches.

Services

Class 35 Advertising; business management; business administration; office functions.

Class 36 Insurance; financial affairs; monetary affairs; real estate affairs.

Class 37 Building construction; repair; installation services.

Class 38 Telecommunications.

Class 39 Transport; packaging and storage of goods; travel arrangement.

Class 40 Treatment of materials.

Class 41 Education; providing of training; entertainment; sporting and cultural activities.

Class 42 Scientific and technological services and research and design relating thereto; industrial analysis and research services; design and development of computer hardware and software; legal services.

Class 43 Services for providing food and drink; temporary accommodation.

Class 44 Medical services; veterinary services; hygienic and beauty care for human beings or animals; agriculture, horticulture, and forestry services.

Class 45 Personal and social services rendered by others to meet the needs of individuals; security services for the protection of property and individuals.

EXAMPLE OF A COEXISTENCE AGREEMENT TO RESOLVE A TRADEMARK DISPUTE IN MULTIPLE COUNTRIES

This Agreement is made the [*insert day*] day of [*insert month*] 200[ ], by and between X [*state individual's name and citizenship or corporate name and state of corporation*] located at [*insert address*] ("X") and Y [*state individual's name and citizenship or corporate name and state of corporation*], located at [*insert address*] ("Y").

WHEREAS, X has adopted and used the mark [*insert trademark*] in good faith in various countries of the world and primarily in respect of [*insert description of goods or services*]; and

WHEREAS, Y has adopted and used the mark [*insert trademark*] in good faith in various countries of the world primarily in relation to [*insert description of goods or services*]; and

WHEREAS X has objected to Y's registration of [*insert trademark*] in [*insert country*]; and

WHEREAS Y has objected to X's registration of [*insert trademark*] in [*insert country*]; and

WHEREAS, both parties desire to develop and pursue their respective businesses and to avoid any confusion between their respective marks;

NOW THEREFORE, in consideration of the mutual covenants contained herein and for other good and valuable consideration, the receipt and sufficiency of which is mutually acknowledged, the parties agree as follows:

1. Y consents to the use and registration by X of the trademark [*insert X's trademark*] in [*insert United States and its territories or recite the countries to which this applies*] in relation to [*insert specific goods/services*];

2. X consents to the use and registration by Y of the trademark [*insert Y's trademark*] in [*insert United States and its territories or recite the countries to which this applies*] in relation to [*insert specific goods/services*];

3. The parties believe that their respective goods/services can coexist in the marketplace without confusion. Further, while the parties recognize that they may from time to time adopt different logos, artwork, and other indicia associated with their respective marks, they agree that they will endeavor to consistently use different logos, imagery, presentations, typestyles, and/or other indicia of origin with their respective marks. The parties do not need permission from one another to change typestyles, logo presentations, artwork and other imagery and indicia of origin associated with their respective marks. In the event of substantial actual confusion, however, the parties will endeavor to avoid and reduce such confusion, as shall be reasonable.

4. X will withdraw any oppositions filed against Y's trademark applications within fourteen days following the signing of this Agreement;

5. Y will withdraw any oppositions filed against X's trademark applications within fourteen days following the signing of this Agreement;

6. The parties will assist one another in obtaining registrations for their respective marks for their respective goods/services of interest as set out in paragraphs 1 and 2 respectively above and will provide each other with such additional written documents as reasonably necessary;

7. This Agreement will be binding upon and inure to the benefit of the parties, their affiliates, successors, and assigns. The term affiliate means any corporation, partnership, licensee, or other entity that owns or controls, is owned or controlled by, or is under the common control with, a party to this Agreement;

8. This Agreement does not constitute either party hereto the agent of the other party or create a partnership, joint venture, or similar relationship between the parties;

9. This Agreement shall commence on the date first above written and shall continue without limit of period.

X's Signature
[*By (if a corporation)*]:
[*Name/Title (if a corporation)*]:
Date:

Y's Signature
[*By (if a corporation)*]:
[*Name/Title (if a corporation)*]:
Date:

EXAMPLE OF AN ASSIGNMENT OF COPYRIGHT

This agreement is made this [*insert date*] day of [*insert month*] 200[ ], between [*insert name of person, their country of citizenship and year of birth or company assigning copyright*] (hereinafter known as Assignor) and [*insert name of person/company to whom copyright is being given*] (hereinafter known as Assignee).

This will confirm our understanding with respect to the copyright ownership in respect of the various materials that have been prepared by Assignor attached as Exhibit "A" (hereinafter the "Works") [*ensure you include in the Exhibit a full description of all the relevant copyright works to be assigned and copies of the same*] and the assignment of such ownership to Assignee.

In consideration of the appropriate payment for such assistance and other valuable consideration paid by Assignee to Assignor (and receipt of which is hereby acknowledged by the Assignor), Assignor hereby transfers, conveys and assigns to Assignee all copyright and rights in the nature of copyright or rights under copyright in the Works throughout the world for the original terms of copyright and all renewals and extensions thereof, together with all rights to sue for previous and subsequent infringements of copyright and to recover damages in respect of all such acts of infringement of the copyright in the Works, together with the right to make such kinds of uses and adaptations thereof and changes therein as Assignee in its sole discretion may determine.

TIP

MORAL RIGHTS

If this is a work of visual art and the assignee does not want the assignor to assert any moral rights, the assignee might try to get the assignor to waive moral rights with langugage such as, "Assignor hereby waives the right to enforce any and all moral rights in the work". This is fairly common practice in the US.

The Assignor shall execute any and all further instruments and otherwise cooperate and assist Assignee with any and all actions that may be necessary or reasonably requested to effectuate this assignment and transfer and maintain the continued protection of each Work subject hereto.

Signed this [*insert date*] day of [*insert month*] 200[ ]

Assignor Name:

Assignee Name:

EXAMPLE OF A NONDISCLOSURE AGREEMENT/ CONFIDENTIALITY AGREEMENT

To: [*insert name of person/company receiving the confidential information*] ("the Recipient")

In consideration of [*insert name of person/company disclosing the confidential information*] ("the Originator") disclosing certain information relating to the project ("the Information"), the Recipient undertakes:

1. to use the Information solely for the purpose of commenting or advising on the [*insert description/name of project*] project;

2. to keep the Information confidential until the Information comes into the public domain through no act or default on the part of the Recipient or the Recipient's agents or employees;

3. not to disclose the Information directly or indirectly to any person unless prior written consent is obtained from the Originator to discuss the Information with the Recipient's other employees or agents;

4. to keep the Information and all information generated by the Recipient based thereon, separate from all documents and other information of the Recipient and under the Recipient's effective control at its usual place of business;

5. not to make any copy, note or record of the Information without the prior written consent of the Originator and that all copies, notes and

records of the Information made by the Recipient will be clearly marked "Strictly Confidential";

6. to return the Information and all copies, notes and records of the Information to the Originator immediately upon request and expunge all Information from any computer, word processor or other device containing the Information;

7. the restrictions herein provided shall not apply with respect to Information otherwise known by the Recipient, or Information that was or becomes part of the public domain without breach of this Agreement by the Recipient or which was independently developed by the Recipient or is required to be disclosed under law, provided the Recipient notifies the Originator prior to making any such disclosure;

8. that it acknowledges and agrees that because of the unique nature of the Information, the Originator will suffer irreparable harm in the event that the Recipient fails to comply with its obligations hereunder and that monetary damages will be inadequate to compensate the Originator for a breach of this Agreement. Accordingly, the Recipient agrees that the Originator will have the right to seek immediate injunctive relief to enforce the Recipient's obligations contained herein.

Signed by: [*insert name of the Recipient*]
On behalf of: [*insert name of company*]
Date: [*insert date*]

NOTES

NOTES

NOTES

NOTES

index

E

F

G

H

I

N

O

P

PUBLISHER'S ACKNOWLEDGMENTS

Dorling Kindersley would like to thank Fran Vargo for picture research and Romaine Werblow in the DK picture library. Thanks also go to Patricia Carroll for copyediting and indexing, and Corinne Asghar, Ann Baggaley, Hannah Charlick, and Tara Woolnough for proofreading and editorial assistance.

Many thanks also to all the organizations that gave permission to reproduce their intellectual property: Apple Computer Inc., Chloe International, the Coca-Cola Company, Dyson Ltd, Eastman Kodak Company, H.J. Heinz Company Ltd, the Jean-Paul Gaultier Company, KFC Corporation, the Michelin Tyre Public Limited Company, Officine Panerai, Penguin Books Ltd, the Shaolin Temple, Shell Oil Company, Starcknetwork, and Transport for London.

AUTHORS' ACKNOWLEDGMENTS

This book would not have become a reality without the help of many participants. Our sincere thanks to the friends and colleagues who provided us with helpful suggestions and comments.

Fred wishes to thank Johann Rupert for his ceaseless encouragement and inspiration for this project and support for all creative endeavors. Larry would like to thank the partners and associates of Fross, Zelnick, Lehrman and Zissu who have enthusiastically embraced this book and to his clients for all the interesting issues and opportunities they have presented over the years.

Together we wish to thank Ron Lehrman, Roger Zissu, David Weild III, David Ehrlich, Susan Douglas, Barbara Solomon, David Donoghue, David Greenbaum, Charles Weigell, Gir Choksi, Karen Lehnert, Kelly Demaret, and Najat Mishalanie for their insightful comments, observations, input, and many hours of fun-filled discourse. We also deeply appreciate the contributions of Yves Istel and Kathleen Begala. Thanks to Natasha Mostert and Jim Stanton, two highly creative individuals, for their enormous inspiration.

Finally, we would like to extend a special thanks to eagle-eyed publisher, Jackie Douglas, and editor, Elizabeth Watson, and the entire team at Dorling Kindersley for their patience, meticulous attention, and superb contributions.

ABOUT THE AUTHORS

Fred Mostert is Chief Intellectual Property Counsel of the Richemont Group of Companies in London, which includes Cartier, Montblanc, and Chloe. He is Past President of the International Trademark Association and a Guest Professor at Peking University. Fred is a member of the New York Bar, is a solicitor of England and Wales, and holds a masters degree from Columbia University School of Law. He was in private practice with Shearman and Sterling and Fross Zelnick Lehrman & Zissu before moving to London.

Fred has assisted designers, chefs, fitness trainers, opera singers, computer scientists, architects, corporate finance specialists, doctors, and bankers on a *pro bono* basis. He has also counseled celebrities and public figures, including President Mandela, Sylvester Stallone, Boris Becker, Stella McCartney, and the Shaolin Monks.

Fred has written widely on intellectual property and is principal author and editor of the book *Famous and Well-Known Marks – An International Analysis.* His writings have been cited in judgments in the United States Federal Court and he received a World Leaders' European Award for "Best Achievement in Intellectual Property Management". He was a member of the World Intellectual Property Organization's Panel of Experts on the Internet Domain Name Process and currently serves on various advisory boards including the Art Science Research Laboratory in New York.

Fred's big passion in life is a love of good food and he masquerades as a food writer on the side. Practising what he preaches, he has filed a trademark application for his *nom de plume*: The Truffleman.

Larry Apolzon is a partner at Fross Zelnick Lehrman & Zissu in Manhattan, a boutique firm of trademark and copyright lawyers whose roster of clients comprises a veritable "Who's Who" of global brand owners and entrepreneurs. He is a member of both the New York and Illinois Bars and is also registered to practice before the US Patent and Trademark Office.

Larry's client base includes luxury groups, entertainment companies, celebrities, musical artists, major and independent record labels, and entrepreneurs from around the world, who all seek to protect their intellectual property in the United States. He is regularly involved in the selection, protection, and enforcement of new brands and he oversees numerous trademark portfolios. Drawing on his patent background, he helped expand Fross Zelnick's practice to include design patent work and has handled hundreds of design patent applications for a broad range of consumer products and designer articles.

Larry is originally from Lorain, Ohio, and graduated *cum laude* from Tufts University with a Bachelor of Science in Chemical Engineering. After briefly attending the School of Engineering at Cornell University, where he realized he enjoyed a course in environmental law far more than studying equations and operating distillation columns, he attended the School of Law at Case Western Reserve University, where he took every intellectual property course available. At the beginning of his career, he practiced in Chicago with a primary focus on patent litigation, before moving to Manhattan in 1988 and joining what is now Fross Zelnick.

Much of Larry's spare time is spent working in his garden and, with the completion of this book, he is contemplating digging into plant patents.

PICTURE CREDITS

The publisher would like to thank the following for their kind permission to reproduce their photographs: (Key: ***a***-above; ***b***-below/bottom; ***c***-centre; ***l***-left; ***r***-right; ***t***-top)
2 Masterfile: Gary Rhijnsburger. **4** courtesy the Michelin Tyre Public Limited Company: (***cl***). Science Photo Library: Eye of Science (***cr***). courtesy the Shaolin Temple: (***r***). © 1998 Shell Oil Company. All rights reserved: (***l***). **5** courtesy of Apple. Apple and the Apple logo are trademarks of Apple Computer Inc., registered in the US and other countries: (***l***). Corbis: Jutta Klee (***cl***). Science & Society Picture Library: (***cr***). Underground map reproduced with kind permission of Transport for London. Based on the original artwork by Harry Beck, 1933: (***r***). **6** akg-images: Erich Lessing (***l***). courtesy Chloe International: (***c***). **7** courtesy Dyson Ltd: (***c***). courtesy Penguin Books Ltd: (***r***). **8** Masterfile: Douglas E. Walker. **10** Getty Images: Taxi/Geoff Brightling. **12** Masterfile: Paul Eekhoff. **13** Getty Images: Stone/ Lori Adamski Peek. **14-15** Corbis: Daniel Aubry NYC. **16** Corbis: James Noble. **17** Getty Images: Hans Pfletschinger. **18** courtesy of Apple. Apple and the Apple logo are trademarks of Apple Computer Inc., registered in the US and other countries. **19** reproduced courtesy Eastman Kodak Company, trademark and copyright owner: (***t***). courtesy Penguin Books Ltd: (***b***). courtesy the Shaolin Temple: (***c***). **20** © 1998 Shell Oil Company. All rights reserved. Shell trademark reproduced by permission of Shell Brands International AG. **21** courtesy H.J. Heinz Company Limited: (***t***). courtesy the Michelin Tyre Public Limited Company: (***b***). **22** Underground map reproduced with kind permission of Transport for London. Based on the original artwork by Harry Beck, 1933. **23** akg-images: (***br***); Erich Lessing (***l***). **24** Corbis: Jutta Klee. **25** Bridgeman Art Library: Cameo Corner, London, UK (***t***). Getty Images: Michael Buckner (***b***). **26** Corbis: Visuals Unlimited (***t***). **26-27** DK Images: Matthew Ward (***b***). **27** courtesy Chloe International: (***t***). **28** courtesy of Apple. Apple and the Apple logo are trademarks of Apple Computer Inc., registered in the US and other countries. **29** courtesy The Jean-Paul Gaultier Company: (***r***). Juicy Salif, Alessi 1990-1991, courtesy Starcknetwork: (***l***). **30** Science Photo Library: Christian Darkin. **31** courtesy of the Coca-Cola Company. **32** DK Images: Tony Souter. **33** courtesy H.J. Heinz Company Limited: (***r***). © KFC Corporation. All rights reserved: (***l***). **35** Getty Images: David Livingston (***tr***). Science & Society Picture Library: (***l***). Science Photo Library: Eye of Science (***b***). **36** SuperStock: Hugh Burden. **37** DK Images: Dave King (***tr***). courtesy Dyson Ltd: (***b***). Science Photo Library: Gusto (***tl***). **40-41** © 2006 Officine Panerai. **42** The Advertising Archives: courtesy of the Coca-Cola Company. **43** The Advertising Archives: courtesy of the Coca-Cola Company. **44** Photolibrary: Phototake Inc/ David Bishop. **45** Photolibrary: Workbook, Inc/Bernstein Photography Darryl. **47** Photolibrary: Workbook, Inc/Marc Simon. **49** SuperStock: Brand X. **103** Corbis: Alan Schein Photography **121** Getty Images: Photographer's Choice/Bix Burkhart. **153** Getty Images: Taxi/BP. **165** Getty Images: Taxi/Phil Cawley. **207** DK Images: Dave King. **233** Ace Photo Agency: John Matchett. **257** Corbis: David Aubrey
All other images © Dorling Kindersley. For further information see: www.dkimages.com